To Mars and Beyond, Fast!

How Plasma Propulsion Will Revolutionize Space Exploration

Franklin Chang Díaz and Erik Seedhouse

To Mars and Beyond, Fast!

How Plasma Propulsion Will Revolutionize Space Exploration

 Springer

Published in association with
Praxis Publishing
Chichester, UK

Franklin Chang Díaz
Chairman and CEO
Ad Astra Rocket Company
Webster, Texas
USA

Erik Seedhouse
Assistant Professor, Commercial Space Operations
Embry-Riddle Aeronautical University
Daytona Beach, Florida
USA

SPRINGER-PRAXIS BOOKS IN SPACE EXPLORATION

Springer Praxis Books
ISBN 978-3-319-22917-1   ISBN 978-3-319-22918-8   (eBook)
DOI 10.1007/978-3-319-22918-8

Library of Congress Control Number: 2017936894

Cover design: Jim Wilkie
Project Editor: Michael D. Shayler

Printed on acid-free paper

This Springer imprint is published by Springer Nature
The registered company is Springer International Publishing AG
The registered company address is: Gewerbestrasse 11, 6330 Cham, Switzerland

Contents

Acknowledgements

Dr. Chang Díaz would like to acknowledge the valuable inputs to the narrative by his beloved wife, Dr. Peggy M. Chang who, for 35 years, has witnessed and supported the commitment of her husband to the VASIMR® project. Her inputs, having lived alongside the long struggle, add a human dimension to the narrative. The authors are also indebted to Dr. Jared P. Squire, Dr. Mark D. Carter and Dr. Timothy W. Glover, all members of the VASIMR® team during the early NASA years, for their valuable contributions to preserving technical accuracy, and to Dr. Stan Milora, Dr. Kim Molvig, Dr. Ronald Davidson (RIP) and others who contributed to the accuracy of the text in some areas where the passage of time had blurred the memory.

In writing this book, the authors have been fortunate to have had five reviewers who made such positive comments concerning the content of this publication. They are also grateful to Maury Solomon at Springer and to Clive Horwood and his team at Praxis for guiding this book through the publication process. The authors also gratefully acknowledge all those who gave permission to use many of the images in this book. The authors also express their deep appreciation to Mike Shayler, whose attention to detail and patience greatly facilitated the publication of this book and to Jim Wilkie for creating yet another striking cover. Thanks Jim!

Some of the images in this book are taken from the authors' personal collections. While they have been enhanced as far as possible, the quality of their reproduction may not necessarily be up to current standards due to the original source material. However, their inclusion is important for illustrating the narrative.

As with many disruptive innovations, the development of the VASIMR® engine has been a long journey, filled with triumphs and setbacks. The story, recounted in these chapters, stands as testimony to the dedication, perseverance and vision of many individuals who, over so many years, supported the project and contributed to the development of the physics foundations of the engine, and later, to the integration of the required technologies to make it viable. No one gets anywhere without someone else's help and the VASIMR® team is certainly no exception. To those who lent a helping hand along our journey, we gratefully dedicate this book.

About the Authors

Dr. Franklin R. Chang Díaz
Chairman and CEO, Ad Astra Rocket Company

Franklin Chang Díaz was born April 5, 1950, in San José, Costa Rica, to the late Mr. Ramón A. Chang Morales and Mrs. María Eugenia Díaz Romero. At the age of 18, having completed his secondary education at Colegio de La Salle in Costa Rica, he left his family for the United States to pursue his dream of becoming a rocket scientist and an astronaut.

Arriving in Hartford Connecticut in the fall of 1968 with $50 dollars in his pocket and speaking no English, he stayed with relatives, enrolled at Hartford Public High School where he learned English and graduated again in the spring of 1969. That year, he also earned a scholarship to the University of Connecticut.

While his formal college training led him to a BS in Mechanical Engineering, his four years as a student research assistant at the university's physics laboratories provided him with his early skills as an experimental physicist. Engineering and physics were his passion but also the correct skill mix for his chosen career in space. However, two important events affected his path after graduation: the early cancellation of the Apollo Moon program, which left thousands of space engineers out of work, eliminating opportunities in that field and the global energy crisis, resulting from the 1973 oil embargo by the Organization of Petroleum Exporting Countries (OPEC). The latter provided a boost to new research in energy.

Confident that things would ultimately change at NASA, he entered graduate school at MIT in the field of plasma physics and controlled fusion. His research involved him heavily in the US Controlled Thermonuclear Fusion Program, managed then by the US Atomic Energy Commission. His doctoral thesis studied the conceptual design and operation of future reactors, capable of harnessing fusion power. He received his doctorate degree in 1977 and in that same year, he became a US citizen.

After MIT, Dr. Chang Díaz joined the technical staff of the Charles Stark Draper Laboratory in Cambridge, MA, where he continued his research in fusion. In that year, the Space Shuttle *Enterprise* made its first successful atmospheric test flight and re-energized the moribund US Space Program. Following this success, in 1977, NASA issued a nationwide call for a new group of astronauts for the Space Shuttle Program. In addition to US citizenship and in contrast to previous announcements in the 1960s, the qualification requirements also included an advanced scientific degree. Dr. Chang Díaz was ready.

Rejected on his first application to the Astronaut Program in 1977, he tried again in a second call in 1979. This time, he successfully became one of the 19 astronaut candidates selected by NASA in May 1980, from a pool of more than 3,000 applicants. He was the first naturalized citizen from Latin America to be chosen.

While undergoing astronaut training, Dr. Chang Díaz fulfilled flight support roles at the Johnson (JSC) and Kennedy (KSC) Space Centers and served as capsule communicator (CAPCOM) in Houston's Mission Control. In 1985, he led the astronaut shuttle support team at the Kennedy Space Center. During his training, Dr. Chang Díaz logged over 1,800 hours of atmospheric flight time, including 1,500 hours in high performance jet aircraft.

Dr. Chang Díaz achieved his dream of space flight on January 12, 1986, on board the Space Shuttle *Columbia* on mission STS 61-C. The 6-day mission deployed the SATCOM KU satellite and conducted multiple scientific experiments. After 96 orbits of the Earth, *Columbia* made a successful night landing at Edwards Air Force Base in California's Mojave Desert.

After a nearly 3-year hiatus, following the *Challenger* disaster of January 28, 1986, Dr. Chang Díaz flew a (world) record 6 more space missions, which contributed to major US space accomplishments, including the successful deployment of the Galileo spacecraft to Jupiter, the operation of the Alpha Magnetic Spectrometer, a major international particle physics experiment, the first and last missions of the US-Russian Shuttle-MIR Program and, on three separate space walks, totaling more than 19 hours outside the spacecraft, where Dr. Chang Díaz led the installation of major components of the International Space Station (ISS) and conducted critical repairs on the Canadian ISS Robotic Arm. In his seven space missions, Dr. Chang Díaz logged over 1,600 hours in space.

Alongside his astronaut duties, Dr. Chang Díaz continued his research in applied plasma physics, investigating applications to rocket propulsion. His 1979 concept of a plasma rocket became the VASIMR® plasma engine, embodied in 3 NASA patents to his name. In 1994, he founded and directed the Advanced Space Propulsion Laboratory (ASPL) at the Johnson Space Center, where he managed a multicenter research team to develop this propulsion technology.

On July 8, 2005, after 25 years of government service, Dr. Chang Díaz retired from NASA to continue his work on the VASIMR® through the private sector. He is founder and current Chairman and CEO of Ad Astra Rocket Company, www.adastrarocket.com, a US private firm based in Houston, Texas, where the VASIMR® engine is being brought to space flight readiness in partnership with NASA. The company is also developing clean energy applications and hydrogen technology at its subsidiary in Guanacaste, Costa Rica.

Dr. Chang Díaz serves on the Board of Directors of Cummins Inc., a global power leader headquartered in Columbus, Indiana, and EARTH University, an international sustainable development educational institution in Costa Rica. He also leads the "Strategy for the XXI Century" http://www.estrategia.cr/, a master plan, aimed to transform Costa Rica into a fully developed nation by the year 2050.

In 1986, Dr. Chang Díaz received The Liberty Medal from President Ronald Reagan at the Statue of Liberty Centennial Celebration in New York City. He is a four-time recipient of NASA's Distinguished Service Medal, the agency's highest honor and was inducted in the US Astronaut Hall of Fame on May 4, 2012. He holds many honorary doctorates from universities in the United States and Latin America and has continued to serve in academia as an Adjunct Professor of Physics at Rice University and the University of Houston. He is married to the former Peggy Marguerite Stafford of Alexandria, Louisiana, and has four daughters: Jean Elizabeth (b. 1973) Sonia Rosa (1978), Lidia Aurora (1988), and Miranda Karina (1995). He enjoys music, flying, and scuba diving. His mother, brothers, and sisters still reside in Costa Rica.

PUBLISHED AUTOBIOGRAPHIES

Dr. Chang Díaz has published two autobiographies:

"*Los Primeros Años*" http://www.adastrarocket.com/BookCover.jpg (ISBN 978-9968-47-133-6), written in Spanish, covers his early childhood and adolescence, growing up in the 1950s and 1960s in Venezuela and Costa Rica where he forms his dreams of space exploration.

"*Dream's Journey*" http://www.adastrarocket.com/BookCover2.jpg (ISBN 978-0-692-33042-5), written in English, sees Dr. Chang Díaz embark on a journey to that dream, alone, as an 18-year-old immigrant, with $50 dollars in his pocket and a one-way ticket to the Land of Opportunity. His American journey unfolds against the backdrop of the tumultuous 1970s and takes him through a decade of adventure and discovery to the pinnacle of scientific achievement.

These books are available by writing to: corporate@adastrarocket.com

April, 2017

Dr. Erik Seedhouse
Assistant Professor, Commercial Space Operations, Embry-Riddle
Aeronautical University

Erik Seedhouse is a fully-trained commercial suborbital astronaut. After completing his first degree he joined the 2nd Battalion the Parachute Regiment. During his time in the 'Paras', Erik spent six months in Belize, where he was trained in the art of jungle warfare. Later, he spent several months learning the intricacies of desert warfare in Cyprus. He made more than 30 jumps from a Hercules C130 aircraft, performed more than 200 abseils and fired more helicopter light anti-tank weapons than he cares to remember!

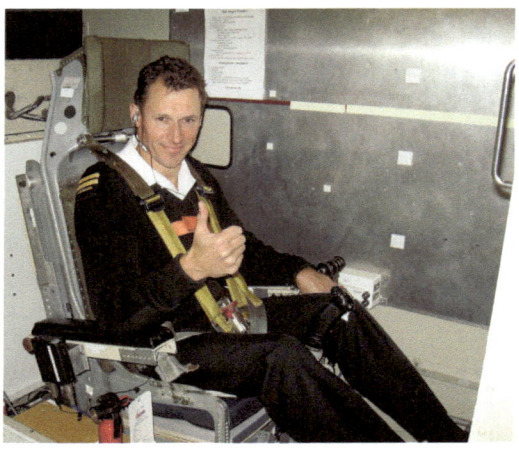

Credit: Chris Townson

Upon returning to academia, the author embarked upon a Master's degree, which he supported by winning prize money in 100km running races. After placing third in the World 100km Championships in 1992, Erik turned to ultra-distance triathlon, winning the World Endurance Triathlon Championships in 1995 and 1996. For good measure, he won the World Double Ironman Championships in 1995 and the infamous Decatriathlon, an event requiring competitors to swim 38km, cycle 1800km, and run 422km. Non-stop!

In 1996, Erik pursued his PhD at the German Space Agency's Institute for Space Medicine. While studying, he found time to win Ultraman Hawai'i and the European Ultraman Championships, as well as completing Race Across America. Due to his success as the world's leading ultra-distance triathlete, Erik was featured in dozens of magazine and television interviews. In 1997 GQ magazine nominated him as the 'Fittest Man in the World'.

In 1999 Erik took a research job at Simon Fraser University. In 2005 the author worked as an astronaut training consultant for Bigelow Aerospace. Between 2008 and 2013 he served as Director of Canada's manned centrifuge and hypobaric operations. In 2009 he was one of the final 30 candidates in the Canadian Space Agency's Astronaut Recruitment Campaign. Erik has a dream job as an assistant professor at Embry-Riddle Aeronautical University in Daytona Beach, Florida. In his spare time, he works as an astronaut instructor for Project PoSSUM, an occasional film consultant to Hollywood, a professional speaker, a triathlon coach and an author. 'To Mars and Beyond, Fast' is his 26th book. When not enjoying the sun and rocket launches on Florida's Space Coast, he divides his time between his second home in Sandefjord and Waikoloa.

Foreword

This book is an incredible story of tenacity, patience and persistence on the part of a young man born in San José, Costa Rica who decided at the age of 7 that he needed to come to the United States to become an astronaut. He was insistent in his conversations with his father, who was equally insistent that he get back to his studies and finish high school if he was to have any hope of travelling to the U.S. to begin his quest. Though not a part of this book, knowing a little bit of the story of the early life of Franklin Ramón Chang Díaz makes it much easier to understand how a single human being could withstand decades of spotty – sometimes zero – support for his dream of creating a rocket engine that would eventually make travel throughout our solar system at unimaginable speeds possible.

Despite his father's initial skepticism and his reluctance to encourage Franklin's dream of moving to the U.S. to pursue his astronaut career, when Franklin reached the age of 17, his father finally gave in and approved of his son's proposed plan to travel to Connecticut to pursue his dream, staying with distant relatives who were willing to take him in temporarily. Speaking no English, Franklin came to Hartford, CT, to finish high school. He had only a one-way ticket to the U.S., $50 in cash and his father's advice: "I send you off with a one-way ticket, because a two-way ticket will tempt you to use it when the going gets tough, and it will. You will fight better this way, but if you must give up the fight, write to me and I will get you back to Costa Rica…" Undaunted by the challenges of his new home country, Franklin taught himself English, graduated with honors from the University of Connecticut and went on to study for his Doctor of Science in Plasma Physics at the Massachusetts Institute of Technology (MIT). It was at MIT that he began his decades-long pursuit of an advanced plasma rocket that would enable space travel at incredibly fast speeds. It is at this point that the book opens.

In what he terms the "Nautilus Paradigm," Franklin fully understood how the U.S. Navy, under the leadership of Admiral Hyman Rickover, developed nuclear-powered propulsion systems to power submarines that would revolutionize transportation on the oceans by allowing a submarine to submerge and travel under the north polar ice cap. It was his belief that applying this paradigm to space travel could revolutionize humanity's ability to "… move from the … Earth-Moon environment … to the deep space interplanetary realm," as he states in the opening chapter of this work.

From his very early days of study at MIT in the 1970s, Franklin was very much aware of challenging impediments to the development of the electric propulsion concept known as VASIMR® (Variable Specific Impulse Magnetoplasma Rocket), but he would not be deterred.

Franklin and I met and eventually became very close friends when we both checked into the NASA Astronaut Selection Process at the Johnson Space Center in Houston, TX in February 1980. As our group – very heavily made up of military test pilots – waited at the old Clear Lake City Airport for pick up, we decided to introduce ourselves around. The test pilots went first, bragging about their backgrounds in various types of high perfor-mance aircraft. Franklin was the last to speak and, barely lifting his head as he spoke very softly with his noticeable Hispanic accent, he said: "My name is Franklin Chang Díaz and I am a plasma physicist." Out of my ignorance, and not intending to be funny at all, I asked him: "Do you work with blood?" I still remember how he looked at me in disbelief as if wondering: "What kind of buffoon is this guy?" After he and I were selected in that second class of Space Shuttle Astronauts, that experience would serve Franklin well about a year after our selection, when he became the first in our class to go on national TV to talk about our training. Franklin was invited to come on the David Letterman Show and he was elated. We all warned him that Letterman was a comedian and that he should not expect any serious conversation during the show. So as not to disappoint, Letterman's very impressive and gracious introduction ended with a very simple question to Franklin: "Do you work with blood?" Franklin laughed it off and launched right into a very down-to-Earth, incredibly clear explanation of the VASIMR® rocket engine.

Working in the Astronaut Office, with its very military style of operational orientation, Franklin faced a clash of cultures as he searched for opportunities to exercise a little of the academic flexibility of the life of a researcher. As he describes it in his third chapter, he found the Astronaut Office to be a workplace led by test pilots steeped in the "military tradition versus the need for a dose of disciplinary diversity." Rather early in our time in the Astronaut Office, I traveled with him to Princeton University to meet and talk with some of his peers involved in early plasma propulsion research. There, we saw an early plasma engine firing, and I began to become a believer in Franklin's dream.

During my fourteen years in the Astronaut Office, I was privileged to fly with Franklin on two Space Shuttle missions – our first and our fourth (which would be my last). I gained increasing respect and admiration for his tenacity and patience in working to help people understand the concept of VASIMR® propulsion, building its scientific credibility with a first-rate research team and addressing the naysayers with hard, peer-reviewed, experi-mental data. I continued to follow his progress after leaving the Astronaut Office and returning to the Marine Corps. Franklin remained undaunted and undeterred by the dis-couraging environment around NASA and JSC, and he and his team finally decided to leave government service and go out on their own. In 2005, he was finally able to found a small company, Ad Astra Technologies Inc. (later Ad Astra Rocket Company), where he would be joined by his small band of young pioneers who shared his belief in the potential for VASIMR® to become a game-changing form of in-space propulsion. Over the follow-ing decade, Ad Astra went on to raise sufficient private investment to prove the remaining physics unknowns and bring the VASIMR® engine to a high level of technological maturity.

Our professional paths would again cross during my tenure as the NASA Administrator in the Obama Administration. I decided to push for, and provide, at least minimal funding to support a search for game-changing in-space propulsion and other systems to support our Journey to Mars efforts, through a competitive process we called the Next Step Technology Exploration Partnerships (NextSTEP) Broad Area Announcement. This provided an opening for Franklin and Ad Astra to compete for NASA funding to advance the ground testing of the VASIMR® engine, as a critical step towards competing for an actual space flight for flight testing the rocket. Ad Astra was selected as one of the winning concepts and was funded for a 3-year, $9 million contract to conduct a long-duration, high-power test of an upgraded version of their VX-200 VASIMR® prototype.

As I write this foreword, Franklin and the Ad Astra team are already performing initial firings of their new engine, the VASIMR® VX-200SS rocket, in preparation for a 100-hour test that they hope will lead to space and the commercial deployment of the technology as primary propulsion for efficient and economical high-power solar-electric space trucks. Later, as we build our human path to the depths of the solar system, a lunar surface test of the rocket in a human-tended lab with multi-megawatt power systems will test the VASIMR® engine that will enable "the Nautilus Paradigm," and take humans to Mars and points beyond.

Though there is still much challenging work to be done for Ad Astra and Dr. Chang Díaz, my money is on their successful ground test and ultimate in-flight use to greatly reduce the transit time of humans to Mars. In an article in *ARS Technica* on February 22, 2017, Eric Berger wrote: *"Truth be told, the plume does not look impressive at all. And yet the engine firing within the vacuum chamber is potentially revolutionary for two simple reasons: first, unlike gas-guzzling conventional rocket engines, it requires little fuel. And second, this engine might one day push spacecraft to velocities sufficient enough to open the Solar System to human exploration."*

As Franklin says in closing out the final chapter of this book, humanity's serious pursuit of human journeys to Mars and other destinations in our solar system will require the cooperation of multiple nations of the world, and a robust exploration program will require the development of advanced nuclear-electric power and propulsion. I have been privileged and honored to have had the opportunity to witness Dr. Franklin Ramón Chang Díaz and his team work diligently against all odds for the past almost 40 years now to bring this vital propulsion technology into reality. Like NASA, he has worked his entire adult life to turn science fiction into science fact and make the impossible possible.

Godspeed, Franklin!

Charles F. Bolden Jr.
Maj. Gen. USMC (Ret.)
12th NASA Administrator
Pilot, STS-61C and STS-31
Commander, STS-45 and STS-60
Alexandria, VA
February 22, 2017

Preface

I had entertained the idea of writing a book about the development of the VASIMR® engine for many years but, somehow, the proper timing never quite arrived; that is, until Erik Seedhouse contacted me with a proposal to jointly undertake the project. He was an experienced writer and had been researching the topic of human space travel for years. I immediately accepted. Originally, the concept was to feature the technology as a means to accomplish ultrafast missions to Mars and beyond; however, while the VASIMR® team considers this as the ultimate application of the technology, we felt strongly that tying the feasibility of fast missions to Mars and beyond solely to the propulsion system would trivialize the myriad of other technologies that must be brought to bear on the success of such missions.

Nonetheless, aware of the strength of the VASIMR® contribution to helping solve the space transportation problem, and of our intimate familiarity with the technology, we chose to focus on its development, staying true to the facts and the hard experimental data along its long historical path. The historical path is also useful to show how non-technical forces often drive the development of a disruptive technology. In the case of VASIMR®, the segregation of plasma physics groups in electric propulsion and magnetic fusion gave rise to the struggle to bring about a convergence of these two cultures, along with that of traditional chemical rocket scientists. Many misconceptions were engendered along the project's nearly 40-year journey, primarily from quick and biased snapshots, by many who were skeptical of VASIMR®, which were never updated and became stale over time. It is also our goal here to dispel or clarify these misconceptions with hard and well-vetted scientific data. We present the evolution of the technology, from its most basic principles and earliest conceptualization, to the high technology readiness, high-power system undergoing tests today. The VASIMR® is being developed by Ad Astra Rocket Company as a high-power electric propulsion system for multiple users; from solar-electric cislunar robotic cargo tugs to nuclear-electric fast human transports. For fast human transport in deep space, however, nuclear-electric is the option of choice. We make this case, as the "Nautilus Paradigm," at the beginning of the book and present a sample mission at the very end. To all of our readers, we hope you enjoy reading this book as much as we have enjoyed writing it.

Franklin R. Chang Díaz

1

The Nautilus paradigm

On August 1, 1958, the *USS Nautilus*, the first nuclear powered submarine, dove from a point off the north coast of Alaska in the North Pacific and surfaced four days later near Greenland in the North Atlantic. Diving under the north polar cap, the "nuclear-electric ship" achieved a feat that no other vessel of its time was capable of and forever changed the strategic balance of sea power.

The transportation breakthrough took place rather quietly, but its impact had profound repercussions which resonate to this day. The development of nuclear power for naval transportation, particularly submarines, occurred very quickly after the dawn of the nuclear age. This profound paradigm shift took less than two decades from the day Enrico Fermi and his team at the Metallurgical Laboratory of the University of Chicago achieved the first controlled nuclear chain reaction, on December 2nd, 1942. That historic feat was demonstrated in a graphite structure, called Chicago Pile 1 (CP1), housing a number of channels filled with uranium oxide. The experiment was conducted in a converted squash court, located under the stadium bleachers at the university's Stagg Field. By 1948, Argonne's Naval Reactor Division had been formed, at one of several US nuclear research facilities spawned by the Manhattan Project, and six years later the *Nautilus* made its maiden sea voyage under nuclear power.

Since its inception in the mid-1950s, naval nuclear power has been a remarkable success story. Power plants in nuclear submarines have had an exemplary service record and modern versions remain basically unchanged from the early design pioneered by the Argonne National Laboratory and later, under the leadership of Admiral Hyman Rickover, by the Bettis Atomic Power Laboratory of the Westinghouse Electric Corporation. Initial testing of the *Nautilus* nuclear-electric propulsion system took place in an earlier version of the shipborne power plant, called the S1W, at the Naval Reactors Facility of the Idaho National Engineering Laboratory (INEL) in eastern Idaho.

Nautilus was powered by a Westinghouse (S2W) pressurized water reactor, fueled by enriched uranium 235 capable of generating 13,400 HP (10 MW) of mechanical power for propulsion. The heat energy from the reactor was transferred to a primary water cooling loop, which also acted as a neutron moderator. The primary loop transferred its heat through a heat exchanger to a secondary loop, which generated steam to drive steam turbines, which in turn generated propulsion and electricity for the ship. Naval reactors are

© Springer International Publishing Switzerland 2017
F. Chang Díaz, E. Seedhouse, *To Mars and Beyond, Fast!*, Springer Praxis Books,
DOI 10.1007/978-3-319-22918-8_1

1.1 The nuclear powered submarine *Nautilus* changed the paradigm of sea transportation

extremely rugged and capable of operating reliably in extremely demanding conditions. Very effective materials engineering and quality controls have been conducted to ensure that corrosion and other material failures are kept in check over years of operation under high temperature and pressure. Radiation exposure levels for personnel in a nuclear submarine are extremely low.

A NAUTILUS FOR SPACE

A "Nautilus paradigm" is required in space for humans to achieve truly robust and sustainable deep space travel: the capability to move from the relatively benign Earth-Moon environment – requiring only conventional chemical propulsion – to the deep space interplanetary realm, which, as in the *Nautilus*, will require high power nuclear-electric propulsion. Yet, since the 1980s, the US (and indeed the world's) investment in nuclear space power research has been paltry at best. Such long-term neglect has created a major deficiency in the technology portfolio needed to carry out a credible, long-term program of human space exploration.

This predicament stems, in part, from the general anti-nuclear sentiment that permeated the world after the Three Mile Island and Chernobyl accidents and, more recently, the natural catastrophe in Fukushima, Japan. Other contributing factors, in the US, are the result of opaque governmental responsibility boundaries between the Department of Energy (DoE) and the National Aeronautics and Space Administration (NASA). These two entities remain largely separate in their respective missions. While the latter is the designated steward of America's space program, the former remains the developer and keeper of the nation's nuclear power technology. In the absence of a higher-level mandate and a suitable coordinating entity, such mission separation hinders the highly integrated technological machinery that must lead an effective space nuclear power program. Other nuclear-capable nations have not done any better. Therefore, the global scarcity of nuclear know-how is a major threat to our future success as a space-faring civilization.

The lessons of the US Naval Nuclear Propulsion Program are clear and compelling. From its early days in the 1950s, the program has remained a comprehensive, fully integrated, cradle-to-grave technology organization, responsible for the research, design, development, testing, operation, maintenance and disposal of naval nuclear propulsion plants. Its extraordinary record speaks for itself: over 150 million miles traveled under nuclear power – more than the average distance between Earth and Mars – and 6,500 reactor-years of accident-free operation. Nuclear submarines are so well shielded that, during a two-month patrol, submarine plant operators receive less radiation from the reactor than they would have received from the normal environmental background while on shore leave.

Another important element of the Naval Nuclear Propulsion Program is its strong tradition of partnership between the private sector – which began in 1949 with Westinghouse and General Electric – and the nation's nuclear research facilities, particularly the Oak Ridge National Laboratory (ORNL), for the most advanced research and nuclear expertise. These partnerships were, however, aligned under the strong centralized leadership headed by Admiral Hyman G. Rickover. Such a robust triangular structure, thriving on discipline and excellence, is needed today in space nuclear-electric propulsion.

The task of developing nuclear-electric propulsion does not need to be viewed as strictly US-centric, but rather it may be a multinational effort by nuclear-capable countries including the US. A close precedent is the ongoing International Tokamak Experimental Reactor (ITER) Project, a multinational effort to build the first demonstration nuclear fusion power plant for terrestrial use. The project, currently under construction in Cadarache, France, is being pursued by several of the world's nuclear-capable countries, including India, Japan, Russia, China, the US, South Korea and the member nations of Europe's EURATOM organization. While the ITER Project has not achieved the same level of leadership and fiscal discipline as the Naval Reactors Program, it stands as a model of international collaboration in a far more complex scientific and engineering undertaking, one whose implementation is arguably more difficult than the construction of the International Space Station (ISS). Indeed, the development of nuclear-electric space propulsion does not need to reach such a high level of multinational diversity, but a long-term commitment by one or more nuclear-capable nations will be necessary to achieve success.

1.2 Admiral Hyman G. Rickover

Nuclear-electric propulsion (NEP) is a "game-changer" and, given sufficient development resources, its full potential could be achieved in time to support deep space human exploration in a sustainable way. Given the inherent limitations of chemical and solar-electric propulsion, it would be difficult to fathom a long-term human presence in deep space without a well-developed nuclear-electric propulsion and power technology. Still, the nuclear theme continues to conjure up controversy, mostly rooted in misconceptions about the dangers to public safety and nuclear proliferation.

Practical commercial nuclear-electric power has been available on Earth since the 1950s and today provides a substantial fraction of the planet's electricity. The process employed in nuclear reactors is called *nuclear fission*, in which nuclei of heavy elements such as uranium[1] are split by subatomic particles called neutrons. A neutron can act as a nuclear "wood splitter," lodging itself into the uranium nucleus and ultimately stressing it sufficiently to break it apart. The nuclear breakup produces more neutrons that go on to split neighboring nuclei, creating a cascade or chain reaction. Besides additional neutrons, the breakups produce chunks of the original nuclei, called fission fragments, which, along with the neutrons, fly off at very high velocities and collide with neighboring atoms, producing a great deal of heat. The heat energy is absorbed by a coolant, which in a heat cycle produces mechanical work to spin an electric power generator that ultimately delivers

---

[1] Other fuels, such as plutonium and thorium, are also available.

electricity to the user. The reactor coolant is often plain water, but gases or more exotic heat transfer media, such as molten salts and some metals, are also used in some designs.

One of the key safety issues in the operation of the reactor is the control of the rate at which the nuclei are being split, or "fissioned," by the neutrons. If the rate is too fast, the reactor overheats, leading to a potential "thermal runaway," also known as a meltdown. If the rate is too slow the reaction dies out. Regulating the reaction between these two opposing extremes is done by controlling the neutron population in the nuclear core. Certain materials act as neutron reflectors that keep the population from scattering away from the core, thus enhancing the reaction rate. Other materials act as neutron absorbers that decrease the neutron population and hence reduce the rate. Control rods made out of boron, cadmium or hafnium, themselves effective neutron absorbers, are mechanically inserted into, or retracted from, the reactor core to control the reaction rate. To shut down the reactor, the rods are fully inserted to rapidly reduce the neutron population and hence the reaction rate.

Several considerations are important regarding human exposure to radiation near the reactor. In the immediate vicinity of the active core, humans must be shielded from the escaping neutrons. This is typically done with graphite shields or water, as the hydrogen in the water is very effective in slowing down the high energy neutrons. There are, however, two other immediate hazards. The fission process also generates energetic electromagnetic waves, known as gamma rays, which are lethal and are largely unaffected by the water. The fission fragments are also radioactive, emitting additional gamma rays, neutrons or other charged particles, which can be harmful if unchecked. Moreover, the fission fragments – elements like strontium, cesium and iodine – remain radioactive for a period of time, eventually decaying to more stable elements but in some cases taking hundreds of years to do so. They must, therefore, be properly contained within the core to avoid radioactive contamination. High energy gamma rays must be stopped with high density metals, such as tungsten and lead, and these shields add significantly to the weight of the reactor core.

NUCLEAR-THERMAL OR NUCLEAR-ELECTRIC?

There are two ways of utilizing the power of nuclear fission for space propulsion: nuclear-thermal and nuclear-electric. In the first approach, the heat of the nuclear pile is simply transferred to a working fluid, typically gaseous hydrogen, which is then expanded and accelerated in a conventional rocket nozzle to provide rocket thrust. In this way, nuclear-thermal rockets (NTR) can reach exhaust velocities nearly twice that of a conventional chemical engine, but are ultimately limited to that level of performance by materials constraints associated with the high temperatures of the exhaust gases. In the 1960s, the United States conducted the Nuclear Energy Rocket Vehicle Applications (NERVA) Program, which demonstrated a nuclear-thermal rocket with nearly 900 seconds in specific impulse[2], a key metric of rocket performance which we shall discuss later in this

---

[2] Specific Impulse (I_{sp}) is a key rocket performance metric. It is simply the exhaust velocity in m/sec, divided by the acceleration of gravity at sea level, 9.8 m/sec^2. It has the units of seconds and its significance in rocket engineering will be described in more detail later in the book, but we provide it here for the reader's convenience.

book. This level of performance is greater by a factor of two than the best chemical rocket, even today. While these results were impressive, pushing the technology much beyond those numbers is not considered practical. In the 1970s, this realization, combined with safety concerns associated with radioactive contamination, led to the project's ultimate cancellation.

The nuclear-electric approach, on the other hand, has no such limitations. In this scheme, the energy from nuclear fission is converted to electricity, which is then used to turn a gas into plasma – a soup of charged particles, positive ions and negative electrons – and accelerate its component particles electrically to provide useful thrust. Most of the thrust in these rockets is provided by the positive ions, which are the more massive of the two, hence the term "ion engine." However, the term "plasma rocket" is more accurate, as the exhaust is actually a plasma, a mixture of an equal number of negative electrons and positive ions. Positive and negative particles must always flow out of the ship together, to prevent the spacecraft building an undesirable negative electric charge which would attract the ions back to the craft, making the rocket unable to provide any thrust at all. Ion propulsion and plasma propulsion are thus interchangeable terms; ion engines are plasma rockets and vice-versa. In all cases, plasma rockets can achieve much higher specific impulse than their chemical or nuclear-thermal cousins.

Microscopically, plasmas are electrically charged fluids, composed of nearly equal numbers of ions and electrons. The ions are chosen over the electrons for acceleration because they are much more massive and, at the same velocity, can carry more momentum. Different electric propulsion technologies use different ion acceleration methods. In the traditional ion engine, the ions are accelerated by DC electric fields imposed by grid electrodes immersed in the plasma. An external neutralizer gun sprays electrons into the accelerated ion stream to produce a neutral plasma jet. Hall thrusters are variants of the ion engine that can reach higher densities in the exhaust jet, by replacing the accelerating grid electrode with a stationary electron cloud held in place by a localized magnetic field. They, too, must neutralize the ion stream with a neutralizer gun, however. In the VASIMR® engine[3] on the other hand, the ions are accelerated by electromagnetic waves in a guiding magnetic field, completely eliminating the need for electrodes. No neutralizer gun is required, as both ions and electrons flow together and exit the rocket at equal rates.

Barring some exotic laboratory exceptions, plasmas are, by nature, very hot. Typical laboratory plasmas can be tens of thousands of degrees; therefore, confining and guiding them in a material duct to make a rocket is a challenge. One solution is simply to keep the plasma density low enough so the particles, while hot, are less numerous and the total power delivered to the wall remains within acceptable limits. This solution imposes an undesirable geometric drawback: to increase the power of the rocket, the size of the engine has to grow accordingly in order to increase the plasma volume without increasing its density.

A more desirable approach is to insulate the plasma from nearby structures by means of a non-material duct; a force field of the appropriate shape and strength. In this way, plasma temperatures and densities well beyond the melting point of materials can be

---

[3]The term VASIMR® stands for Variable Specific Impulse Magnetoplasma Rocket. VASIMR® is a registered trademark of the Ad Astra Rocket Company.

achieved, which in turn increases the power density of which the rocket is capable. These physics-driven parameters will be discussed later in this book. In general, traditional ion engines, governed by space charge and materials limitations, have the lowest plasma density and hence the lowest power density. Hall thrusters can reach higher densities by replacing the accelerating grid electrode with the stationary electron cloud held in place by a localized magnetic field. Even higher densities are attainable in the VASIMR® engine, where power is delivered by electromagnetic waves, thus removing density limitations and completely eliminating the need for electrodes. In VASIMR® systems, power densities of several MW/m^2 are achievable.

ELECTRIC PROPULSION: A PATH FROM SOLAR TO NUCLEAR

Given an electric engine such as the VASIMR®, able to process so much power, we return to discussing the particulars of the electric power source needed to drive it. The VASIMR® engine is insensitive to its source of electric power, and indeed the Ad Astra Rocket Company envisions its earliest commercial applications not as nuclear, but as solar-electric, operating in the Earth-Moon environment at power levels of hundreds of kW. Solar-electric technology has matured to the point where such capability is technologically viable and actually extremely attractive from the standpoint of in-space transportation economics. Nonetheless, the VASIMR® engine also scales very well to multi-megawatts and thus its ultimate deployment in the nuclear-electric realm in support of human deep space exploration is the focus of this book.

A nuclear reactor is a heat engine[4], not that different from a coal or gas furnace. The heat produced must be turned into useful electricity by a power conversion system; for example, a steam turbine driving an electric generator. When one examines the power generated from the heat engine and follows the conversion of this power into mechanical work and finally into electricity, the useful output turns out to be about 30-40 percent; the rest is waste heat that must be dissipated. As heat engines go, a conversion efficiency of 35 percent is fairly typical with today's technology.

Higher efficiencies are clearly desirable. Unfortunately, the typical conversion of heat to mechanical and electrical energy is governed by the laws of thermodynamics, which impose limits to the attainable efficiency. In their 2011 study on multi-megawatt nuclear-electric space power, Dr. Ronald Litchford from NASA Marshall Space Flight Center and Dr. Nahiburo Harada from the Nagaoka University of Technology in Japan, described an advanced Magneto Hydrodynamic (MHD) power system that achieves a 55 percent power conversion efficiency on a net electric power output of 2.76 MW. Their nuclear-electric architecture makes use of direct energy conversion of a fast-moving, weakly ionized, gaseous working fluid that transfers its energy to an electric field in a magnetic expander. The electric field drives the voltage source in an electric circuit, which in turn drives an electric current to produce useful work.

---

[4]A heat engine is a system that produces mechanical work from heat. The mechanical work can be used directly for locomotion or to drive machinery, or indirectly by producing electricity which is then used in multiple applications.

Such direct energy convertors became popular in the mid 1970s when the energy crisis of 1973 drove major advances in electrical power generation. Unfortunately, the resurgence of cheap oil in the 1980s indefinitely postponed the implementation of such advances into the mainstream. Nuclear power also stalled, following the accidents at Three Mile Island and Chernobyl. Technically speaking, MHD power conversion was not a panacea in the 1970s, as there were many difficulties associated with the cost of these systems, including the need for superconducting magnets – expensive and complex at the time – for the magnetic expander, and the "seeding" of the high speed gases with chemically ionizing compounds that produced the charged particles needed to transfer the energy to the electric field, a process which was also technically challenging. Moreover, chemical "seeding" was environmentally questionable due to chemical pollution concerns, which diminished the attractiveness of these early embodiments of the technology.

In the 1960s, space nuclear-electric propulsion technology did not fare much better. Several drawbacks, including the large mass of the power conversion system and radiators required to shed the waste heat, discouraged its maturation. The important mass considerations associated with nuclear-electric propulsion are usually summarized into one single parameter, called the system "α" (alpha), defined as the ratio of the total mass of the combined system (power and propulsion) to the total electric power. As the α of the system is reduced, the attractiveness of NEP over all-chemical and nuclear-thermal architectures becomes evident. Present state-of-the-art alpha values hover around 10, but some tantalizing technology concepts have surfaced which could bring this number into the single digits. Sadly, the space nuclear-electric power field has been neglected for many decades and very limited actual technology development has taken place since the 1960s.

Much has changed technologically since the dawn of the 21st Century, however, as major advances in high temperature superconductivity and RF-based ionization technology have opened new options for direct power conversion systems, which could reduce alpha and bring high power nuclear-electric propulsion back to prominence. For example, in Litchford and Harada's study, the overall mass of an advanced power plant was found to be between 2 and 3 kg/kW, with alpha values potentially lower than unity for systems above ten megawatts. These possibilities must be explored in earnest, as they open extraordinary advantages for fast deep space missions under nuclear-electric power.

Another drawback to high power nuclear-electric propulsion has been the lack of a sufficiently mature high power electric rocket engine that would be compact enough to be married to such a low alpha power source. Suitably powerful ion engines, the only mature technology at the time, were too large due to their inherent low power density. Moreover, their high-voltage power processing equipment was too heavy and bulky to be operated reliably at power levels of several megawatts. High power density electric rockets, such as the VASIMR® engine and others currently under early development, are poised to eliminate this problem.

While Ad Astra is not in the business of developing space power sources, the company carefully follows the progress of both the leading space electric power options: solar and nuclear. For its near-term robotic commercial applications, Ad Astra foresees (within 5-10 years) high delta-v VASIMR® flights maneuvering payloads in the "low Earth orbit" (LEO) to "geostationary Earth orbit" (GEO) regions of space, powered by solar-electric arrays. Combined with state-of-the-art support and deployment mechanisms, these arrays should be able to provide power (out as far as Mars) at a specific mass in the range of 2-7 kg/kW (the range depending on radiation shielding requirements) – much lower than the best nuclear space power systems developed to date.

Thus, in the near term, using solar-electric power at levels of 100 kW to 1 MW, VASIMR® propulsion could transfer heavy payloads to Mars using only one to four first-generation thrusters in relatively simple engine architectures. By optimizing the ratio of power to total vehicle mass at an appropriate specific impulse, significant cost savings over chemical in-space propulsion can be realized. This application should be attractive for a methodical, cost-effective, long-term plan of Mars exploration, in which infrastructure and supplies are pre-positioned at Mars by slow cargo flights in advance of faster human transits. This is a capability that can be demonstrated first at relatively low power levels in support of robotic exploration, and then grow as space electric power generation improves.

Such improvements point squarely to the "Nautilus Paradigm," the need for advanced nuclear-electric power. In this realm, much remains to be done and development work is a long-term effort that must not be delayed. Ad Astra has explored the scaling of the VASIMR® technology to multi-megawatt engines driven by nuclear-electric power and has conducted interplanetary mission studies of very high power architectures. These studies, discussed in Chapter 10, yield a wide range of fast interplanetary mission options, with one-way trip times to Mars ranging from four months to just over one month, depending on the performance of the nuclear power source (specified in kilograms/kilowatt, kg/kW). It is abundantly clear that the nuclear reactor technology required for such missions is not available today and major advances in reactor design and power conversion will be needed. However, a number of serious research studies have been conducted that point to reactor and power conversion designs that meet the kg/kW ratio required for such a mission. Much remains to be done and closing the door on these possibilities on the basis of the relatively primitive state of our present nuclear space technology would be highly premature.

2

A Fast Track to Deep Space

Plans to put boots on the surface of Mars have been on the drawing board for decades. In the 1960s during the Apollo Era, the public anticipated that astronauts would be visiting Mars in the early 1980s, but in the United States, the euphoria of the first Moon landing faded quickly. The Americans had won the space race and their attention moved to more pressing earthly issues. The Vietnam war was a festering wound that needed urgent attention and the Watergate scandal plunged the nation into a deep reassessment of its core values. In 1973, the nation, along with most of the Western economies, was engulfed in an energy crisis driven by the Arab oil embargo, a crisis that touched all citizens in their most sensitive spot; their pocketbooks. Gasoline and heating oil shortages became common-place, with long lines of thirsty motor vehicles patiently waiting for fuel at filling stations and prices skyrocketing overnight. Faced with all this, fanciful missions to the Red Planet were as far away as the planet itself.

In the intervening decades leading up to the present, global attention has shifted away from space to more pressing issues at home: terrorism, economics, energy and climate change. In the United States, the initial driving force for space exploration – military supremacy in the sky – was greatly diminished by the political collapse of America's only credible space competitor, the Soviet Union. Today, with more than half a million objects orbiting the Earth, space activity has morphed into a more complex ecosystem with a much larger and diverse set of stakeholders, including a growing number of space-faring nations and commercial satellite operators; a business opportunity in a growing $300 billion market. The machinery of global communications, spawned in large measure by the space age and later lubricated by the Internet, has, perhaps unexpectedly, democratized space. As a result, countries like India and China have built domestic space programs and rocket launch capabilities that rival those of the more established players, the US, Russia, Europe and Japan.

The United States has been slow to recognize and adapt to this organic transformation. Cargo and human transport to low Earth orbit (LEO) became technologically mature in the 1990s and needed to be privatized. Yet it was not until the turn of the 21st Century that the nation initiated a Commercial Orbital Transportation System (COTS) program to spur the private sector into providing these services more cost-effectively and efficiently, via Public Private Partnerships (PPPs) which could better leverage public funds. The sustaining cost

© Springer International Publishing Switzerland 2017
F. Chang Díaz, E. Seedhouse, *To Mars and Beyond, Fast!*, Springer Praxis Books,
DOI 10.1007/978-3-319-22918-8_2

of the old paradigm on an $18 billion dollar US civil space budget has been high, leaving very little wiggle room for deep space exploration. It is therefore no surprise that, half a century after humans visited the Moon, the date for a potential landing on the Red Planet has been pushed back to sometime in the 2030s.

Fifty years after Apollo, the problems associated with deep space travel remain as clear and present as they were in the 1960s. Amazingly enough, in the United States, the approach to their resolution also appears to be frozen in time. The technology of deep space transportation has advanced little, due to very low investment in new, "game-changing" systems, such as solar- and nuclear-electric propulsion (SEP and NEP, respectively). In the United States, the main transportation elements for deep human space exploration remain strikingly similar to those of Apollo: a very large chemical rocket and a capsule capable of returning a small human crew back to Earth from a point not much farther away than the Moon. The lion's share of the Mars mission architecture remains in the planning stage – where it has been for decades – and, while its transportation strategy does leave the door open for a nuclear-thermal option, the nuclear-electric approach has not been seriously explored. This is an omission of considerable significance, an arguably naïve and unsustainable strategy which needs to be re-examined, given what humans have learned from half a century of space flight.

The operational challenge of safely transporting humans to Mars and back is fundamentally different from a journey to Earth's Moon. Our Moon orbits the Earth, so missions to the Moon technically never completely leave Earth orbit. The Earth is always at the same distance and, in the event of a major malfunction, conveniently no more than 3-4 days away. The same is not true for a journey to Mars, as both Mars and the Earth orbit the Sun and their relative distance changes constantly – and by a much larger measure – over the course of two years. With current chemical rocket technology, typical one-way transits to Mars can take between 7-9 months, depending on fuel and rocket performance. Upon arrival, because of the long transit, the crew must await more than a year for the opening of the return window. This constraint imposes severe requirements on the reliability and survivability of the crew support infrastructure. There is no argument that such reliability could eventually be achieved with organic refinements of current technology. However, it would be foolish not to examine adjacent space transportation technologies, such as high power SEP and NEP, that could substantially change the operational landscape and enable a more rapid and sustainable Mars exploration program.

To be sure, getting to Mars is not the problem; getting to Mars *fast* is. Thus, the Mars debate centers around two important questions: Should we go to Mars now, or should we focus on developing the transportation technologies that will ensure a robust and sustainable program? On the one hand, one could argue that despite the long journey times inherent with conventional propulsion, Mars can be explored, maybe even colonized, with present technology. To many who wish to go now, the radiation threat associated with the long journey is acceptable; moreover, through the experience of ISS, the human space program has now developed the means to tackle many of the other human health and crew habitability issues associated with the mission. On the other hand, with current transportation technology, orbital mechanics and the sheer length of the flight produces a mission architecture that is operationally fragile. In addition, the ISS research continues to uncover as yet unexplained issues of concern in human physiology associated with long duration space flight.

In the post-Apollo era, the US debate on the journey to Mars has been fueled by these deliberations for many years. It has produced multiple embodiments of Apollo-like programs that have all stalled when confronted with budget realities. To avoid this pitfall, it is important to recognize the new chemistry of space. The US-Soviet dipole of the 1960s no longer exists. The forces driving space exploration are now truly global, commercial, economic and political, and with an increasing number of space-faring nations deeply involved, genuinely multinational. A sustainable human Mars exploration program must reflect all these elements and take advantage of this new paradigm. The exploration and colonization of Mars and other deep space destinations is no longer the business of one or two superpowers, but of all the people of Earth; a fact that could be turned into a major resource multiplier. In addition, rather than being a nationalistic Apollo-like stunt, the journey to Mars should take a more practical route by constructing a multinational scaffolding of technology-based transportation; one whose robustness is based on multiple players with overlapping – and even competing – capabilities and not solely on the nationalistic pride and political will of one nation. Such a construct could generate tangible commercial, scientific and economic dividends along the way, well before a landing on the Red Planet. For example, one could envision more cost-effective space logistics delivery, in-space resource utilization and commercial mining of space natural resources as potential benefits.

A TIME FOR CHANGE

In the last few years, the US space program has begun to address these elements with a renewed emphasis on high power electric propulsion, which could naturally evolve from solar to nuclear power sources. High power electric rockets, such as the VASIMR® engine and variants of the Hall thruster, have reached an advanced technology readiness level (TRL) and are poised to be demonstrated in space soon. These, and others still in early development, could provide the aforementioned scaffolding to Mars, while enabling revenue-generating business opportunities in efficient, low cost space transportation closer to home.

Advanced space transportation development must be a technology continuum, running from the near-term more mature systems to the more speculative ones, but always subject to rigorous, well-qualified scientific vetting and experimental verification. While it would be foolish to dismiss futuristic propulsion concepts, relying on matter-anti matter reactions, thermonuclear fusion and space-time warps, these systems, just like all the others, must respond to rigorous scientific scrutiny. Too distant a visionary outlook can be a detriment to progress, as it distracts attention from the middle ground, where new technologies do accrete into practical systems that could be early precursors to the more futuristic ones but can now be experimentally demonstrated and characterized. In fact, focusing too much on the far future is often a way to keep it from becoming the present. Disruptive technologies not only disrupt technologically but also financially, affecting funding streams to established programs and ultimately people. Therefore, to the established paradigm, it is non-threatening to support advanced technologies as long as they continue to remain in the realm of the future, where funding needs are minimal. This reality is often the reason why the middle ground is generally sparsely populated. The established paradigm resists change by clinging to the purse strings. It is thus important to recognize and address this pitfall.

It is also important to recognize that new technologies, such as high temperature superconductors, plasma engineering, nuclear power, advanced materials and manufacturing – all of which could be relevant to NASA's mission – often originate outside of NASA. Therefore, appropriate mechanisms for integrating these advances must be preserved through strong inter-agency programs and public-private partnerships that foster innovation and creativity while preserving scientific rigor.

High power electric propulsion is a case in point. Its genealogy has roots in the field of gaseous electronics as well as thermonuclear fusion, both of which were peripheral to the early NASA, who mainly focused on chemical propulsion. The space agency did undertake some preliminary incursions in these fields, with the work of Harold Kaufman on a variant of the "duoplasmatron" plasma source that led to the modern ion engine. In the late 1960s and 1970s, the space agency also delved briefly into radio frequency (RF)-heated plasmas and controlled fusion, with its research on the NASA-Lewis Bumpy Torus Experiment at the Lewis Research Center (now the Glenn Research Center at Lewis Field) in Cleveland, Ohio.

Early work on the VASIMR® engine was initiated at the MIT Plasma Science and Fusion Center (PSFC), as a non-fusion variant of the Tandem Magnetic Mirror fusion concept, with design features borrowed from magnetic divertors present in Tokamak fusion experiments. As we discuss extensively in the chapters that follow, this early work continued for more than a decade before the system was moved to NASA's Johnson Space Center. Another high power electric rocket, the Magneto Plasma Dynamic (MPD) Thruster, was originally developed in the late 1950s and early 1960s as a plasma injector by John Marshall at the Los Alamos National Laboratory and Hannes Alfvén at the Royal Institute of Technology in Stockholm. The device was known as a Marshall Gun and had applications in experimental plasma physics and the early work in controlled fusion. Later development on the thruster variant of this system was carried out primarily at Princeton University's Department of Mechanical Engineering and later at the Jet Propulsion Laboratory (JPL), a university laboratory closely associated with NASA. The pioneering work on the Pulsed Inductive Thruster (PIT) originated at Northrup Grumman, before the research was pursued by the NASA Marshall Space Flight Center (MSFC) in Huntsville, Alabama. Preserving this strong synergy of the space program with academia, national laboratories and private industry is essential to prevent scientific stagnation and technological inbreeding within NASA and to ensure a healthy accretion of new ideas and discoveries that are also scientifically well vetted and will enable advanced propulsion systems to eventually reach the mainstream.

CHARTING THE GLOBAL PATH TO SPACE EXPLORATION

The foregoing discussion should not project the impression that chemical propulsion is obsolete. Much to the contrary; for the foreseeable future, chemical rockets will remain the best and only practical means of leaving and landing on a planet. The technology of these systems has evolved over many decades to an exquisite level of refinement. The next generation of chemical rockets will enhance reusability and reliability and also reduce cost, all of which are necessary to deliver the optimal scaffolding for deep space exploration.

Although chemical rocket propulsion is a mature technology, and thus is well poised for cost reduction by the stimulation of strong commercial competition on a global scale, its widespread use has been hindered by international restrictions stemming from its military applications. That the rocket was introduced to the world as an instrument of mass destruction is sad and unfortunate. Perhaps humanity has matured sufficiently in the 21st Century, to recognize its value as an instrument of our survival. One would hope that unnecessary international restrictions will gradually disappear as more nations acquire rocket know-how or develop it indigenously. While orbital-capable rockets in the 1960s were the sole purview of the United States and the Soviet Union, nearly a dozen nations have this capability today, a number that is sure to grow quickly if a competitive, revenue-generating global market promotes it. The science and technology of rocket propulsion is today sufficiently well understood, to the point that nations with technologically well-educated populations should be able to master low Earth orbit space flight with moderate capital investment. As interesting examples, private companies such as SpaceX, Blue Origin and XCOR, in rather short timespans, have developed their own indigenous rocket technologies.

Just as space becomes truly multinational, the traditional role of the private sector in space is also beginning to change, from that of a mere government contractor to that of a government partner. This is a healthy evolution that fosters competition and will tend to increase efficiency and reduce both costs and technology maturation time. Humanity is increasingly dependent on a space infrastructure that supports global communications, provides situational awareness to people all over the planet and monitors the state of its life support system. The maintenance of these assets represents a $300 billion business with a lot of room to grow. Such growth can help finance a healthy and sustainable expansion of humanity into space.

In charting humanity's route to deep space, a great deal of debate has ensued regarding the role of the Moon and whether or not our natural satellite should be the next logical destination. It clearly is. We are, in fact, fortunate to have such an excellent proving ground so close to hand for the technologies that will enable astronauts to venture far into the solar system and learn to work efficiently on another world. As a convenient site for testing multi-megawatt plasma engines, the Moon is second to none, and Ad Astra Rocket Company intends to build a rocket test facility on its surface for long-duration tests of multi-megawatt VASIMR® engines under solar- or nuclear-electric power. These tests would become prohibitively expensive and complex in Earth-bound vacuum chambers or free flying spacecraft. Yet they will be required to certify these high power electric engines for long duration operation at full power.

We have spent a great deal of time talking about going to Mars, but looking through the optics of Apollo and conventional propulsion. In the meantime, other technologies have matured that could fundamentally change the architecture of the mission. In high power electric propulsion, these include high temperature superconductors, compact and high power solid-state RF technology, advanced materials and manufacturing, solar-electric power generation, nearly zero boil-off cryogenic propellant storage, advanced controls, and many others. These technologies should have been folded into the space transportation equation years ago. Unfortunately, this process was inhibited partly by the overly "operational" mind-set permeating much of the Space Shuttle Program in the 1980s and 1990s. During this period, the US space agency's long-term strategy for the nation's deep space transportation became fragmented and dispersed; nuclear-electric space propulsion and power has been explored with a great deal of institutional fear. This unfortunate condition may finally be abating with the new Space Technology Mission Directorate (STMD), recently established at NASA. With a sufficiently visionary and enlightened leadership, this centralized technology coordination entity could have the wherewithal to bring about the space equivalent of the "Nautilus Paradigm."

Finally, the focus on Mars has obscured the fact that several other solar system destinations also beckon humanity: the moons of Jupiter and Saturn, where water is now known to be abundant, may provide even more tantalizing opportunities for the existence of life and, with fast and robust space transportation, human explorers may be quickly drawn to these destinations. Journeys to these more distant worlds will indeed be long. They will be well beyond the capabilities of chemical or nuclear-thermal rockets and will require fully autonomous nuclear-electric ships with advanced life support systems and a nearly unlimited range, resulting from the long-lived nuclear fuel and the use of local resources. A power-rich nuclear-electric architecture will bring about these capabilities. The development of nuclear-electric propulsion and power is an urgent need that should not be postponed in the haste of reaching Mars, as without it, humanity will not be able to truly free itself from the bonds of Earth.

In the chapters that follow, we shall describe the history of the VASIMR® engine, from its genesis in the early 1980s to its present highly advanced technology maturation stage. There are many important lessons in this historical journey, but one that stands out is that the implementation of new ideas requires not only a sound technical base, but also a strong dose of persistence.

3

Early VASIMR® Development

All rockets work on the principle of action and reaction: "to every action, there is an equal and opposite reaction." By this principle, the rocket moves by expelling material at high velocity in the direction opposite to the rocket's motion. A common misunderstanding is that a rocket's exhaust "pushes" on its surroundings to propel itself. This is definitely not the case. In fact, friction from its surroundings actually slows down a rocket. Traveling in a vacuum is best. With nothing to slow it down, as long as its fuel lasts, a rocket can accelerate to very high velocity, making interplanetary trips not only possible, but also fast.

But how long can the fuel last? To answer this question, we note that the thrust of a rocket – the force imparted to the ship by the rocket exhaust – is simply the product of the exhaust velocity (relative to the ship) and the propellant mass flow. This means that the same thrust can be achieved by either ejecting more material at low velocity, or less at high velocity. Clearly, since propellant must be carried on board the ship, the latter approach is more desirable. Thus, an important requirement for a rocket on an interplanetary mission is to achieve the highest possible exhaust velocity. Propulsion engineers, however, like to use the term specific impulse (I_{sp}), which is the exhaust velocity divided by the acceleration of gravity at sea level, 9.8 m/sec^2. So, for quick estimates, the exhaust velocity is roughly 10 times the I_{sp}. The higher the I_{sp}, the more fuel-efficient the rocket is.

There are, however, no shortcuts in the laws of nature, and increasing the exhaust velocity comes at a price. That price is energy. As the exhaust velocity increases, the energy required to enable the increase grows as the square of that quantity, increasing the power requirements of the rocket exponentially. This is a challenge for an electric rocket, where power is limited as it must be produced onboard from a solar array or a nuclear reactor. Although modern solar arrays have become increasingly powerful, they do not come close to the equivalent power capability of a chemical rocket. They would also be of no practical use in deep space, far away from the Sun. Nuclear reactors, on the other hand, are a clear high power choice, but their technology for space electric propulsion applications is not yet mature.

One way to address the power shortcoming is to reduce the propellant mass flow, a quantity that scales linearly with power. This, in fact, is the current approach in electric propulsion but this choice also comes at a price, and that price is the delivery of a lower thrust. It is true, therefore, that modern electric rockets are, by nature, low-thrust devices

© Springer International Publishing Switzerland 2017
F. Chang Díaz, E. Seedhouse, *To Mars and Beyond, Fast!*, Springer Praxis Books,
DOI 10.1007/978-3-319-22918-8_3

and this constraint has limited their mainstream use. It is also true that these rockets are very efficient in their propellant consumption, but this frugality comes not only as a consequence of their high specific impulse but also due to their power limitation. Abundant space nuclear-electric power could change that paradigm, providing sufficient power to afford significant increases in thrust without sacrificing specific impulse.

Another important issue arises from the properties of available rocket materials. The exhaust velocity relates directly to the temperature of the exhaust gases. Modern chemical rockets have evolved to a highly advanced state over decades of development, with materials and nozzle cooling schemes allowing exhaust temperatures of several thousand degrees C (Celsius). Nonetheless, despite these impressive achievements, the exhaust velocity of these systems is limited to about 5,000 m/sec, which turns them into veritable "gas-guzzlers." To move past this limit, we must alleviate, or altogether remove, the physical constraints imposed by materials. We must also move away from chemistry and explore ways of achieving temperatures well beyond the capability of chemical reactions. Enter the field of plasma physics.

THE REALM OF PLASMA PHYSICS

In the early 1970s, while working on his thesis research in plasma physics, Dr. Franklin Chang Díaz realized that material impediments to containing, ducting and expelling superhot gases were not unique to rockets. In the field of controlled thermonuclear fusion, scientists routinely contained and ducted super-hot gases, called plasmas, using strong magnetic fields, shaped into "non-material enclosures." Because the "enclosures" were "non-material," they were relatively impervious to the high temperature plasmas they contained. The goal of the fusion scientists was – and still is – to achieve thermonuclear ignition of the plasma, creating a "small sun" on Earth that could release abundant amounts of energy, which would be converted to electricity for human use.

That lofty goal, however, has proven to be more difficult than anticipated and has remained a work-in-progress for more than 50 years. Nonetheless, while not yet hot and dense enough for fusion, modern-day plasmas can reach temperatures of millions of degrees C and densities high enough to be applicable to rockets. With a sufficiently powerful electrical power source, such plasma-based rockets can produce major increases in performance over their chemical cousins. Their practical application is now coming of age, due to the convergence of a number of new technologies that enable them to operate efficiently and be built in relatively small, lightweight packages.

In his early conceptualization of the VASIMR®[1], Dr. Chang Díaz recognized that thermonuclear conditions were not needed for a compelling rocket application. He posed the following question: Would it be possible to create a magnetic rocket nozzle that could, like the plasma containers for fusion, be impervious to the temperatures required for useful thrust? If the temperature of the rocket exhaust could be elevated to more than about 10,000 degrees C, the atoms in the gas would begin to shed electrical charges and the gas would gradually become plasma, an electrically conducting fluid that, as in fusion, could be contained by a strong magnetic field away from the material walls of the nozzle.

---

[1] VASIMR® stands for Variable Specific Impulse Magnetoplasma Rocket. VASIMR® is a registered trademark of the Ad Astra Rocket Company.

With such a non-material duct, there would be nothing to melt! Better yet, the hotter the plasma, the more electrically conducting it would be and the greater the grip the field could exert on the plasma. If the temperature were, for example, a million degrees C, the plasma would be magnetically isolated from nearby structures and highly constrained to flow down the magnetic pipe. The field could be produced by a lightweight superconducting electromagnet, with the proper distribution of current windings to appropriately shape the magnetic duct in the form of a magnetic nozzle. Similar to a conventional nozzle, a magnetic nozzle converts the plasma's internal energy into directed flow velocity, producing rocket thrust. Because of the high temperatures, the exhaust velocity under these conditions is extremely high, producing a rocket of unprecedented performance.

Reaching those temperatures, however, is well beyond the capability of any chemical reaction. Fortunately, in the early 1970s, a variety of non-chemical plasma heating techniques were already well developed and available, including high power lasers and particle beams driving strong electrical currents and shocks in the plasma. Also available were radio waves and microwaves for generating high temperature plasmas. All of these methods, of course, require electricity, so VASIMR® became, by its nature, an electric rocket. In contrast to Earth-bound fusion, the technology needed to be compact, efficient, and lightweight enough to fly in space, requirements that immediately eliminated most, but not all, of the above plasma heating methods.

To skeptical observers from the space propulsion community in the early 1980s, the proposed VASIMR® system seemed heavy, complex and inefficient. There were also unknowns regarding some of the controlling physics of the device. For example, the ionization of the propellant was considered to be too energy intensive, rendering the engine inherently inefficient. Also, it was argued that the heating of the plasma to high temperature, envisioned by means of ion cyclotron waves, might inefficiently couple electrical power to the working plasma. The flow dynamics in a conceptual magnetic nozzle were also not well understood and many anticipated that the plasma would be incapable of detaching from the

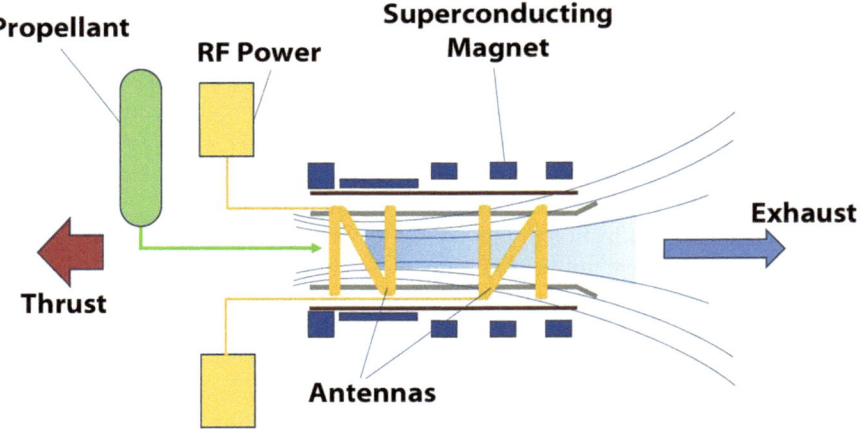

3.1 Simplified schematic of the VASIMR® engine.

expanding field and hence provide no thrust at all. All of these were valid concerns that needed answering. Moreover, beyond the complexities of the rocket itself, another issue loomed ominously; where would the electrical power come from to drive such a rocket?

Spanning more than three decades of research, from its early inception at the Massachusetts Institute of Technology (MIT) in the early 1980s, then later at NASA in the 1990s and at the Ad Astra Rocket Company in the first decade of the 21st Century, Dr. Chang Díaz and the VASIMR® team set out to address all of these issues systematically and thoroughly. In the 1990s, several key experiments began to shed light on the physics of the VASIMR® engine. A number of enabling technologies, which the team had anticipated, actually began to mature. For example, with the development of high-power metal oxide semiconductor field effect transistors (MOSFETS), lightweight, reliable, efficient and compact solid-state RF sources became a reality. In addition, advanced, high-temperature ceramics, highly transparent to RF waves, became suitable wall materials for plasma-facing chambers. High temperature superconductors had also progressed to the point where high magnetic fields could be generated in an efficient, compact and lightweight package, operating at temperatures an order of magnitude greater than conventional superconductors. The higher superconducting temperatures have greatly eased the refrigeration challenge and cost. These advances have enabled the design of a VASIMR® engine which meets and exceeds the requirements for flight. All of these considerations will be addressed in detail in the chapters that follow, but first, a brief discussion of space electric power, a key ingredient of all electric propulsion architectures and not unique to VASIMR®.

SPACE ELECTRIC POWER

The power source plays a critical role in the utility and performance of all electric rockets. Unlike chemical rockets, whose fuel contains the propulsive energy (in the form of chemical bonds), electric rockets must supply the propulsive energy to a propellant[2] from an external electric power source, such as a solar array or a nuclear reactor. In this sense, chemical rockets are said to be "energy limited," whereas electric rockets are "power limited." In the chemical rocket, the amount of energy available is directly proportional to the amount of fuel available. In the electric rocket, on the other hand, the amount of energy available (from the Sun or a nuclear reactor) is virtually unlimited; the only limitation being how much of it can be pumped into the propellant every second. Over a given amount of time, electric rockets are capable of injecting more total energy to each kg of propellant than their chemical cousins – the higher the I_{sp} the more energy the propellant can absorb.

Historically, however, space electric power has been limited by the capability of early space nuclear and solar power technology. But in recent years, solar arrays have improved significantly in their power generation capacity, from a few tens of Watts/m^2 in early technology to hundreds of Watts/m^2 in present arrays. With adequate funding and commitment, space nuclear-electric technology could also greatly advance to the point where acceptably compact and reliable multi-megawatt nuclear-electric space power sources are

---

[2] The term *fuel* is not appropriate for electric rockets as there is no combustion, so *propellant* is used instead.

a reality. High power electric rockets such as VASIMR® will benefit from these developments, as they can process large amounts of power in relatively small packages. The net result, in a transportation sense, is a higher energy trajectory and hence, a faster interplanetary transit. This high power scalability has led many to think that the VASIMR® engine needs a nuclear reactor to operate. Such is not the case. VASIMR® engines can work just as well with solar-electric power and indeed their initial use, in the business of robotic space logistics in cislunar space, will employ solar-generated electricity.

While the availability of large amounts of electric power is important, the optimal management of that power is also essential. As it turns out, for a constant amount of power, thrust and I_{sp} are inversely related and function in a manner similar to the gears in a car. In a high gravity environment, near a planet, the available power is better utilized in the form of thrust, which provides more maneuvering muscle. As the ship moves away, and the grip of the planet's gravity is less, power is better utilized to increase the I_{sp}, resulting in more efficient fuel consumption and higher speed. This "shifting of gears" is called Constant Power Throttling (CPT), since it is done while keeping the total amount of power constant (at maximum). As we shall see in the chapters that follow, the VASIMR® engine has CPT capability and could use it throughout any given mission to optimize fuel consumption.

Abundant electric power is now increasingly recognized as a critical and indispensable component of a robust and sustainable human space exploration program. With regards to electric propulsion, near the Earth-Moon environment and within the inner solar system, the Sun's rays are sufficient to drive high power solar-electric propulsion (SEP) systems up to ~1MW. High efficiency, lightweight solar arrays are being developed, capable of power densities of ~400 W/m². These can be folded in compact packages for cost-effective delivery to space aboard modern chemical launchers. For power levels above one megawatt, suitable for fast human interplanetary transportation, there is little doubt that advanced nuclear-electric power sources are required. This is particularly true as missions move farther from the Sun.

The nuclear reactor technology required for such missions is not available today and major advances in reactor design and power conversion are needed. If payoffs such as enhanced mission capabilities and applications are to be realized, development work must start in earnest. Fast Earth-Mars transits, potentially ranging from four months to just over one month in duration, depending on the performance of the nuclear power source (generally specified in kilograms/kilowatt), are a handsome payoff, worthy of a committed effort by nuclear-capable nations. A small number of research studies have been conducted that point to reactor and power conversion designs that meet the kg/kW ratio required for such advanced missions.

ELECTRIC PROPULSION AND PLASMA ROCKETS

Plasma propulsion was not new in the 1980s. Since its early days, the field has been known by its more prosaic name: electric propulsion. Its genesis was more closely tied to low power applications in gaseous electronics than to controlled fusion. Electric propulsion was studied by rocket pioneers such as Robert Goddard in the early 1900s and Ernst Stuhlinger at the end of World War II. Understandably, from its early origins, the field evolved on a low power diet, where watts instead of kilowatts were generally the power norm because electric power was a scarce resource on early spacecraft. Space power generally came from early solar photovoltaic arrays and, in very special cases, from small and low-power nuclear "batteries" called radioisotope thermoelectric generators (RTGs), at best capable of hundreds of watts. In the late 20th Century, hydrogen-oxygen fuel cells, capable of tens of kilowatts, were employed in human spacecraft such as Project Apollo and the Space Shuttle, which required more robust electrical generating capacity. In the last 20 years, however, solar array technology has improved sufficiently to power the International Space Station (ISS) at power levels of 70-100 kW.

Historically, research funds for electric propulsion have always been dwarfed by expenditures in chemical propulsion. In the modern age of intercontinental ballistic missiles (ICBMs) capable of delivering a nuclear warhead to the enemy's backyard, rocket research by the world's superpowers has been driven by missile design and defense requirements, rather than by the needs of space exploration. Despite these priorities, greater improvements in solar technology are expected, making high power electric propulsion an increasingly competitive option in terms of payload mass fraction. As these technologies mature, electric propulsion is poised to take center stage as humanity heads outbound into the solar system.

The earliest concept for a plasma rocket is the gridded ion engine, where ions are extracted from tenuous plasma discharges and accelerated by a DC electric field. Other variants of these ion accelerators evolved over time, leading to the Hall Effect Thruster (HET), a technology of Russian heritage which has been tested at tens of kilowatts. Ion engines and HETs are ideally suited for satellite "station-keeping" and other low power applications. Commercial versions of these thrusters are indeed playing an important role in the realization of the "all-electric spacecraft." Because of their inherently low power density (see figure 3.2), scaling of these systems to hundreds of kilowatts and to megawatts, as would be needed to support a robust deep space human exploration infrastructure, becomes less practical as the power level increases. High power applications using these systems requires clustering of an increasingly large number of units. Thus, the perceived value of increased engine redundancy is quickly offset by complexity and size.

In the 1970s, Dr. Chang Díaz's approach to an electric plasma rocket departed from these low density plasma discharges. It involved several pieces of fusion technology that he had explored during his PhD thesis at MIT and during his tenure at the Charles Stark Draper Laboratory in Cambridge, Massachusetts. One of these was known as a magnetic divertor, a device designed to magnetically "peel away" plasma from the edge of a thermonuclear reactor core in order to sweep away impurities migrating there from the vessel wall. In a fusion reactor, impurities tend to cool the plasma by stimulating more radiation. For the plasma core to stay hot enough to produce fusion, it is important to keep the plasma very clean.

Unfortunately, the magnetic field, the "fabric" insulating the plasma from nearby materials, is not perfect and some of the hot plasma particles from the thermonuclear core do manage to diffuse through it and bombard the material walls of the vessel, releasing impurities of carbon, tungsten, and other heavy atoms. The magnetic divertor proved a very successful device to eliminate unwanted impurities before they had a chance to penetrate deep into the plasma core. So successful, in fact, that the divertor became a standard design feature in the Tokamak, the most advanced type of fusion experiment of the day.

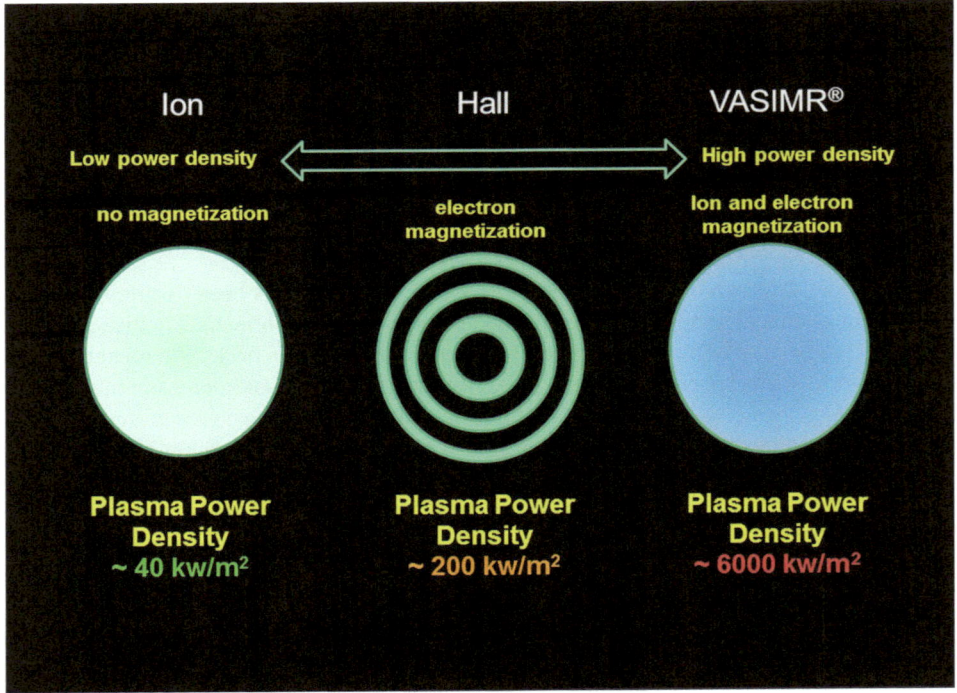

3.2 Typical power density (kW/m2) regimes for three electric propulsion technologies.

But what to do with the million-degree C plasma debris scooped away by the divertor? In most designs, the divertor includes a magnetic channel that guides unwanted plasma to some sort of dump chamber, where it is allowed to impinge directly onto a target material surface and deposit its energy. The surface is cooled to remove large amounts of heat. However, over time, the surface material erodes away and must be replaced. In the late 1970s, as a potential remedy for this problem, Dr. Chang Díaz proposed replacing the material target with a gas "curtain." The gas would have to be moving fast enough to avoid gas diffusion up the plasma stream into the reactor chamber. A high-speed flow was also necessary to carry away the plasma heat.

The fast moving "curtain" was a supersonic jet, a stable gaseous structure for the plasma to impinge upon and deliver its energy. He called the fast moving curtain, appropriately, "The Gas Target Divertor." His interest in divertors led him to other fusion experts working on that technology, such as his colleague at Draper Labs, Dr. Jay L. Fisher, and Dr. Tien Fang "Ted" Yang, of MIT. A small collaboration effort on gas target divertors for fusion began to take form between Draper Labs and MIT. However, while this work was getting underway, Dr. Chang Díaz was already contemplating a variant of the Gas Target Divertor which he theorized could form the basis of a plasma rocket. He called this concept the Hybrid Plume Plasma Rocket, the early precursor to the VASIMR® engine.

The magnetic duct comprising the divertor channel is simply a cylindrical magnetic pipe formed by a set of electrical current rings or coils. These coils are distributed along the length of the device to produce an invisible magnetic pipe with either a smooth or a corrugated topology. If the coils are close together, the magnetic pipe resembles a smooth, straight cylinder threaded through the rings; however, if the coils are sufficiently spaced, the field is no longer straight but tends to bulge between the rings, producing a topology that resembles a string of loosely tied sausages. Dr. Chang Díaz's arrangement of coils produced a magnetic duct with three linked but distinct chambers, serving different but complementary processes. The first chamber, "the ionizer," produced low temperature plasma from a feedstock of neutral gas. The second chamber, "the heater," received this plasma and heated it by radio waves to very high temperatures. The last chamber, "the nozzle," was an open magnetic nozzle, where the heated plasma would naturally acceler-ate in the expanding magnetic field and leave the device to produce rocket thrust.

There was another important capability inherent in the design, which Dr. Chang Díaz had begun to explore: the ability to vary the exhaust velocity (the specific impulse) and the thrust, without changing the power setting of the engine. This "Constant Power Throttling" or CPT, as he called it, was the feature similar to shifting gears in an automobile. If more thrust was required, more of the power would be directed to the "ionizer" and less to the "heater," to make more plasma. A denser but cooler exhaust would result, providing more thrust, albeit consuming more fuel. Alternatively, by shifting more of the power to the "heater," less plasma would be generated, providing less thrust, but the exhaust would be faster and more fuel-efficient.

For high power rockets, this variability is important when moving in the gravitational "hills and valleys" near planets, as well as the "flat terrain" of open interplanetary space. If needed, and in order to enhance the detachment of the plasma from the magnetic field, a co-axial, hypersonic layer of neutral gas could be injected in the nozzle at a shallow angle, in order to form a gas sheath around the plasma jet and prevent significant diver-gence of that plasma jet as the magnetic field expanded past the last coil. Such co-axial jets resemble the flow of the working vapor in a diffusion pump. When sufficiently energized, the plasma easily detaches from the nozzle without the need for a gas sheath. However, for low plasma temperature operation (low I_{sp}, high thrust mode), the extra help can further enhance detachment at low exhaust speeds.

There was another purpose for the coaxial jet, which Dr. Chang Díaz contemplated in his early designs. Mixed with the plasma, the coaxial jet would create a sort of "plasma afterburner" that would provide bursts of extra-high thrust when needed. This was an oper-ationally attractive feature that could provide greater maneuverability to the spacecraft.

Such was the rocket of the future, but in 1980, Dr. Chang Díaz needed to take a pause from scientific exploration to start preparing to fly a less futuristic rocket called the Space Shuttle. On May 31 of that year, he and 18 other Americans, from a pool of more than 3000, were selected by NASA to comprise the 9th group of US astronauts. He and his teammates reported for duty on July 8 to the Lyndon B. Johnson Space Center (JSC) in Houston to begin training to fly on future Shuttle missions. Although he had been a natu-ralized US citizen since 1977, Dr. Chang Díaz's country of origin was Costa Rica and his selection to the program made him the first NASA astronaut from Latin America. Two European "candidates," Claude Nicollier from Switzerland and Wubbo Ockels from the Netherlands, also arrived for training in Houston that summer, raising the number of rook-ies to 21 and giving the program a new international flavor.

A MEETING OF TWO CULTURES

Soon after his arrival at NASA, Dr. Chang Díaz began seeking ways to continue developing his plasma rocket concept, in conjunction with his astronaut training. The work seemed even more relevant now that he was part of America's Space Program. The Astronaut Office, however, was a completely different environment to those he had been accustomed to during his years at MIT and Draper. It was a community of "operators," a culture of test pilots with strong and deeply ingrained military traditions. Despite a new dose of disciplinary diversity, injected by the new breed of young scientist astronauts of which he was a part, the astronaut "operational" ethos was still largely unchanged. This was totally justified and reasonable, given the importance of honing the safe, efficient, and reliable operation of the Space Shuttle. However, the mindset tended to shun scientific research, considering it a distraction incongruent with the duties of an astronaut.

In Dr. Chang Díaz's mind, the astronaut ethos had to evolve from that of a mere operator to that of an explorer and front-line investigator. This was particularly true as space operations involving humans became more scientifically ambitious, complex and diverse. It was clear that an exclusionary choice between "operations" and "science" was neither required nor desirable. Operations and science had to coexist and meld into a hybrid mode that Dr. Chang Díaz called "Operational Science." In the development of his propulsion concept, Dr. Chang Díaz sought to balance his new duties and training as an "operator" of a unique flying machine with a sustained and appropriate level of research.

Soon after settling down in his new office, on the third floor of Building 4 at JSC, Dr. Chang Díaz discussed his plasma propulsion work with his superiors, Chief Astronaut, Captain John W. Young and Flight Operations Director, Mr. George W. S. Abbey. Both of them were supportive and encouraged him to continue the development of the rocket, as long as it did not conflict with his space flight training requirements. After reviewing Dr. Chang Díaz's research plan, Mr. Abbey allocated a small budget of approximately 10,000 dollars for him to conduct his rocket development, under a new element of the astronaut training syllabus called Astronaut Technical Proficiency Training.

This management decision was pivotal in enabling a small seed to grow, albeit in an unlikely field. Historically coming from piloting careers, all NASA astronauts were required to be – and remain – proficient in their flying skills. In selecting astronauts from scientific backgrounds, however, NASA was changing the mix and adding a new skill set to the corps: the ability to conduct interactive scientific research while flying in space. For scientist astronauts, it would not be possible to achieve this requirement without maintaining their proficiency in their respective fields. Technical Proficiency Training became an important ingredient to ensure the fulfilment of a new astronaut duty; namely interacting, in real time, with increasingly complex scientific investigations being conducted in space.

Starting in 1981, Dr. Chang Díaz initiated a small collaborative program in plasma physics with his former Draper partner, Dr. Jay L. Fisher, as well as Dr. Ted Yang at the MIT Plasma Science and Fusion Center (MIT-PSFC). While neither Draper nor the MIT-PSFC were programmatically inclined to detour from their fusion research mission and delve into rocket propulsion, the embryonic work was approved by their management, as it was of a theoretical nature and general enough to provide relevance to both fields. The MIT/Draper institutional link was exactly what Dr. Chang Díaz needed to continue to develop the characteristics of the plasma rocket.

On November 24, 1981, Dr. Chang Díaz described the early VASIMR® concept in his logbook as:

"…a hybrid type plasma device. A linear confined plasma such as in a Tandem mirror with an external axial magnetic field. The thrust is created by feeding the loss cone plasma in the keV temperature range into the core of an expanding gas jet – maybe in an annular nozzle. The result is a plume temperature profile which meets the low temperature wall requirement at the edge but is allowed to increase to the plasma temperature near the plume centerline…"

By 1982, the collaborative work on magnetic divertors between the Astronaut Office and the MIT/Draper team allowed for the Gas Target Divertor to be further characterized. The first self-consistent description of the concept was published by Dr. Chang Díaz and Dr. Fisher, in a 1982 paper entitled "*A Supersonic Gas Target for a Bundle Divertor Plasma* [1]," which appeared in the peer-reviewed Journal of the International Atomic Energy Agency, *Nuclear Fusion*.

At the same time, the conceptualization of the plasma rocket continued and an invention disclosure for a "Hybrid Plume Plasma Rocket" was presented to Mr. Carl McClenny of the Office of Patent Counsel at NASA JSC on August 6, 1982. Several years would pass, however, before two patents were granted, in 1989 and 1990, on the embodiment of the rocket and the method of plasma propulsion respectively. The substantiation of the patent claims on these inventions required a great deal of numerical work, involving advanced plasma codes residing in the supercomputers of the National Magnetic Fusion Energy Computer Center (NMFE-CC) at the Lawrence Livermore National Laboratory (LLNL). Housing several Cray I, II and CDC 7600 mainframes, the NMFE-CC was, at the time, the most powerful computer facility in the world.

In a time well before the advent of the Internet, many scientists were already diligently making use of the NMFE-CC through the Internet's precursor, the Arpanet, a government operated national computer network which linked US National Laboratories and major universities and research centers nationwide. Draper Laboratory was one of these research centers and during his tenure there, Dr. Chang Díaz made extensive use of the Arpanet to obtain remote access to the NMFE's advanced plasma and magnetic field codes to carry out his work in controlled fusion. Time on these machines was a precious resource, made available by the Energy Research and Development Administration (ERDA) – the precursor to the US Department of Energy (DoE) – to many plasma physicists and fusion scientists nationwide.

Unfortunately, the Astronaut Office was not connected to this network. In the Space Shuttle era, the NASA Johnson Space Center had become an "Operations" and training facility, whose role was to train astronauts and fly the Space Shuttle safely and frequently. Some management projections estimated that the Space Shuttle would fly as often as once per week. Accordingly, during the Shuttle build up, the JSC organization became increasingly populated by operations personnel from the various branches of the military. Research activities, which had grown during the Apollo program, were on the decline and were gradually being eliminated. It was no surprise, therefore, that despite having led the human landing on the surface of the Moon a decade earlier, most at JSC had never heard of Arpanet and the virtual world of number-crunching accessible at the NMFE-CC.

To remedy this, Dr. Chang Díaz obtained approval from his NASA management to install a Tektronix terminal in his office, along with a 1200-baud modem and an acoustic

United States Patent [19]

Chang

[11]   Patent Number: **4,815,279**

[45]   Date of Patent: **Mar. 28, 1989**

[54]   **HYBRID PLUME PLASMA ROCKET**

[75]   Inventor: Franklin R. Chang, Webster, Tex.

[73]   Assignee: The United States of America as represented by the National Aeronautics and Space Administration, Washington, D.C.

[21]   Appl. No.:   46,341

[22]   Filed: **Apr. 13, 1987**

Related U.S. Application Data

[63]   Continuation of Ser. No. 781,397, Sep. 27, 1985, abandoned.

[51]   Int. Cl.⁴ F03H 1/00
[52]   U.S. Cl. 60/202; 60/264; 239/265.17
[58]   Field of Search 60/202, 203.1, 204, 60/39, 462, 265, 264, 231; 239/127.3, 265.17

[56]   **References Cited**

U.S. PATENT DOCUMENTS

2,862,099	11/1958	Gage	60/203.1
2,906,858	9/1959	Morton, Jr.	60/203.1
3,005,338	10/1961	Libby et al.	60/265
3,013,384	12/1961	Smith, Jr.	60/203.1
3,119,233	1/1964	Wattendorf et al.	60/202
3,151,449	10/1964	Manson	239/127.3
3,173,248	3/1965	Curtis et al.	60/202
3,239,130	3/1966	Naundorf, Jr.	60/202
3,360,682	12/1967	Moore	60/202
3,520,139	7/1970	Elkind et al.	239/127.3

4,369,920	1/1983	Schmidt	60/265
4,663,932	5/1987	Cox	60/202

FOREIGN PATENT DOCUMENTS

570334   2/1959   Canada 60/264

Primary Examiner—Donald E. Stout
Attorney, Agent, or Firm—Hardie R. Barr; John R. Manning; Edward K. Fein

[57]   **ABSTRACT**

A technique for producing thrust by generating a hybrid plume plasma exhaust is disclosed. A plasma flow is generated and introduced into a nozzle which features one or more inlets positioned to direct a flow of neutral gas about the interior of the nozzle. When such a neutral gas flow is combined with the plasma flow within the nozzle, a hybrid plume is constructed including a flow of hot plasma along the center of the nozzle surrounded by a generally annular flow of neutral gas, with an annular transition region between the pure plasma and the neutral gas. The temperature of the outer gas layer is below that of the pure plasma and generally separates the pure plasma from the interior surfaces of the nozzle. The neutral gas flow both insulates the nozzle walls from the high temperatures of the plasma flow and adds to the mass flow rate of the hybrid exhaust. The rate of flow of neutral gas into the interior of the nozzle may be selectively adjusted to control the thrust and specific impulse of the device.

6 Claims, 6 Drawing Sheets

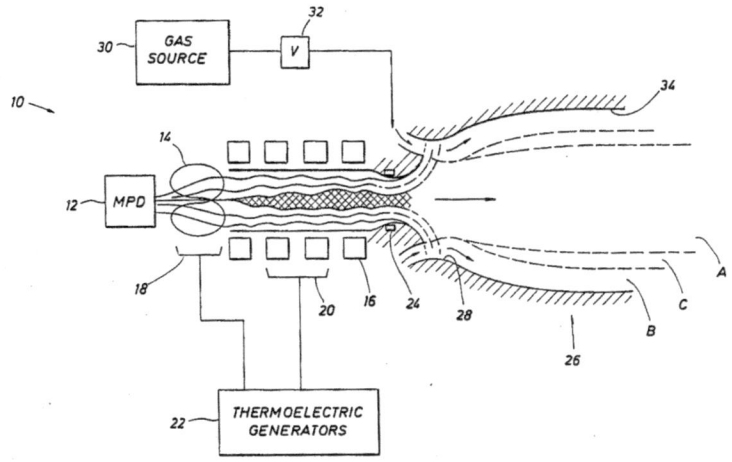

3.3 First US patent addressing the controlling principles of the VASIMR® engine.

coupler[3] to which he could connect his "push-button" phone[4]. Using this equipment, he was able to access the Arpanet through Tymnet, a California-based private computer network linking thousands of mainframes nationwide. A simple local telephone call from his push-button phone at JSC to a downtown Houston node of Tymnet established the bridge to Arpanet and to the NMFE computers at Livermore.

Even for astronauts in the early 1980s, making frequent – and potentially expensive – long-distance calls from their government office phones required some explanation, so it was fortunate that a Tymnet node could be found in the Houston area. Curiously enough, the node was maintained by the oil industry and access to it from a government telephone was a simple local call, which would not trigger management attention. It was an intricate, convoluted, and, at 1200-baud, slow way to reach the awesome computing capability of the US Government supercomputers, but the set up worked and, much to the amazement of his fellow astronauts, the green phosphorous screen of the Tektronix terminal would display detailed 3-D contours of the early VASIMR® engine plasma every few minutes, generated at the NMFE computers in California in response to the commands from Houston.

It was thus that the first virtual research facility for plasma rocket development was established within the "operational" organization of the Astronaut Office. The new capability linked the MIT-PSFC and LLNL with the NASA Johnson Space Center and allowed the slow but interactive evolution of the early design of the VASIMR® engine. At MIT, two PhD graduate students, Scott Peng and Warren Kruger, began their theses in the rocket program, under the supervision of Dr. Yang and Dr. Chang Díaz, who had been named a visiting scientist to the PSFC.

THE ELECTRIC PROPULSION COMMUNITY

It was important, however, to establish a link with the electric propulsion community and begin a difficult process of culture convergence between two somewhat immiscible groups: the high power plasma physicists of magnetic fusion, accustomed to large and powerful machines, and the smaller group of investigators focusing on low power gaseous electronics, relevant to the electric rockets of the day. In the early 1980s, Dr. Chang Díaz was relatively unfamiliar with the technology advances and challenges of electric propulsion. It was clear that the low power available to spacecraft severely limited their capability and mission applications. However, in his astronaut familiarization briefings at NASA Headquarters and at the agency's field centers, he had learned of nuclear-electric power sources of increasingly higher power being developed for deep space exploration. One of these was the SP-100 space nuclear reactor, under development by General Electric with participation from NASA, the Department of Defense and the Department of Energy.

The SP-100 was projected to be capable of delivering 100kW of electric power to a spacecraft and its design was presumed to scale to much higher power. One of its applications was, of course, as a power source for electric rockets. Plans were being drawn to

---

[3] Prior to the advent of Wi-Fi, acoustic couplers were used to acoustically connect a telephone receiver to a modem and establish a computer communication link.

[4] Some rotary dial phones were still in use in the 1980s.

conduct a space test of the device on board the Space Shuttle and Dr. Chang Díaz was assigned to represent the Astronaut Office in this program and work as part of a committee established to oversee its safety. The program, however, lost momentum, particularly after the accident of the Space Shuttle *Challenger* in January of 1986, and was terminated in the early 1990s. Another nuclear reactor program, called Prometheus, was established briefly in the 1990s, using a reactor of Russian design called Topaz, which had some similarity to the SP-100. This program, too, was short-lived.

Despite these uncertainties, research on the early VASIMR® continued through the JSC-MIT collaboration. In 1982, Dr. Chang Díaz and fellow rookie astronaut, USMC Commander Charles F. Bolden Jr., paid a visit to the Electric Propulsion and Plasma Dynamics Laboratory (EPPDyL) at Princeton University, headed by Dr. Robert Jahn. During his graduate student years, Dr. Chang Díaz had visited the nearby Princeton Plasma Physics Laboratory (PPPL), one of the DoE National Laboratories engaged in controlled fusion research. In the mid-1970s, the PPPL was engaged in the construction of the most advanced US fusion experiment at the time, the so-called Tokamak Fusion Test Reactor (TFTR), and Dr. Chang Díaz had been sent there to study its design as it related to his PhD thesis on fusion power systems.

Curiously enough, however, during the 1982 visit to Princeton's EPPDyL, Dr. Chang Díaz became aware that the facility was not part of the larger PPPL, but instead existed somewhat isolated from the fusion community, within the infrastructure of Princeton University's Department of Mechanical Engineering. During their visit, Dr. Chang Díaz and Cdr. Bolden met with Dr. Kenneth Clark, Deputy Director of the EPPDyL, and toured the research facility, which at the time was engaged in the development of the Magnetoplasma Dynamic, or MPD, thruster, a device known as a Marshall Gun plasma injector in the fusion community. Surprisingly enough, there seemed to be little interaction on electric propulsion research between EPPDyL and nearby PPPL, a finding that further buttressed the need for the culture convergence Dr. Chang Díaz was seeking.

The EPPDyL visit was actually very useful to begin a link to the small US electric propulsion community. Shortly after Princeton, and with leads from Dr. Clark, Dr. Chang Díaz made contact with Dr. Robert Vondra, a specialist in small, pulsed plasma thrusters, who was serving as a civilian scientist at the US Air Force's Rocket Propulsion Laboratory (RPL) at Edwards Air Force Base (AFB), in California's Mojave Desert. Dr. Vondra had been working in the field for several years and was familiar with the key players, especially his west coast colleagues at the Jet Propulsion Laboratory (JPL), an institution of the California Institute of Technology heavily funded by NASA. After a phone call from Dr. Chang Díaz, Dr. Vondra arranged a visit to JPL, where Dr. Chang Díaz could brief his VASIMR® concept.

The JPL visit had both negative and positive results. On the negative side, a high power plasma rocket such as VASIMR® was well outside the JPL team's comfort zone of 1-2 kW ion engines, so they reacted with a great deal of skepticism over the practicality of the technology. Power levels of hundreds of kilowatts and higher were considered unrealistic and the team was unfamiliar with high density magnetized plasmas and the rapid evolution of superconducting magnet technology. On the positive side, however, Dr. Vondra became interested in the physics of the VASIMR® engine and in learning more about the rocket's potential. He arranged for Dr. Chang Díaz to provide another briefing to his home team at the RPL at Edwards AFB.

The visit to RPL took place shortly after the JPL briefing and had more positive results. Following the presentation, Dr. Vondra initiated a review of the concept by his colleague, Dr. Leonard Caveny, Director of the Air Force Office of Scientific Research (AFOSR) at Bolling Air Force Base in Arlington, Virginia. The AFOSR had begun investing small sums of research funds in advanced electric propulsion concepts, and a $50,000 proposal from Dr. Chang Díaz to study the VASIMR® engine was accepted and funded. As the first contract ever obtained for the development of the VASIMR® engine, this funding stands today as a historical milestone. The AFOSR contract provided a stable financial vehicle to engage in some limited level of computational work and continue Kruger and Peng's PhD thesis research. These two activities began to define the operational plasma parameter space that needed to be explored by a future experiment.

By 1986, Dr. Chang Díaz had completed his first space flight, STS-61-C, on the Space Shuttle *Columbia*. The flight was a six-day mission to deploy an RCA Satcom KU telecommunications satellite and send it on its way to its parking position in geostationary orbit, 36,000 kilometers above the Earth's equator. The crew was also trained to conduct a number of scientific experiments in medicine and microgravity, as well as a bit of astronomy. By coincidence, Comet Halley was making one of its 75-year-cycle close approaches to the inner solar system and could be photographed from orbit. Technical glitches with the camera prevented the photography, but fortunately the comet was clearly visible with the naked eye, appearing as a fuzzy oversize star low on the horizon. For Dr. Chang Díaz, his first mission into space, albeit short and fairly uncomplicated in its objective, was the most personally impacting. It represented the fulfilment of a life-long dream, which he had pursued from a very young age in a different country; a dream that, unbeknown to him at the time, would repeat itself six more times with ever increasing complexity in the two decades ahead.

The STS-61C crew landed on the Mojave Desert's dry lake bed at Edwards AFB on January 18, 1986. However, the happiness and extraordinary sense of well-being that followed the successful space mission were short-lived. Ten days later, the *Challenger* disaster plunged the space program and the nation into a deep and introspective space recession that lasted nearly three years. The investigation of the accident prompted major changes to safety protocols in Shuttle operations. The long road to recovery ended in September of 1988, when the Space Shuttle *Discovery* launched from the Kennedy Space Center (KSC) in Florida on STS-26 and the space agency finally returned to flight operations. It was a difficult hiatus, during which Dr. Chang Díaz and many of his fellow astronauts became deeply involved in multiple activities associated with the return to flight process, which meant that time for extra-curricular scientific research was significantly reduced. However, as the recovery period came to an end and many of their assigned tasks were completed, crewmembers once again looked forward to flight assignments. Dr. Chang Díaz also resumed his research activities at the MIT-PSFC, in combination with his astronaut training in Houston.

FROM THEORY TO EXPERIMENT

The interaction between Houston and Boston, although greatly reduced during the Shuttle program recovery, never completely ceased, and in 1986, plans to build a small research linear magnetic mirror machine to study the physics of the VASIMR® began to take form. The relationship with AFOSR had continued to grow over the years, to the point where a small experimental program was feasible by combining the funding from AFOSR and NASA's Astronaut Office. Moreover, through the virtual dialogue afforded by the Tymnet-Arpanet connection, Dr. Chang Díaz and his team quickly conceived and developed the architecture of such an experiment. The apparatus would be housed at the MIT-PSFC.

Once again, the need for culture convergence became evident from the start. At MIT, the small and more traditional community of low power electric propulsion, resident primarily in the Department of Aeronautics and Astronautics, was highly skeptical of the new experiment and did not support it. The MIT fusion group, on the other hand, was much more receptive and enthusiastic. The MIT-PSFC group was composed mainly of investigators from the Physics, Nuclear, Electrical and Mechanical Engineering Departments, as well as Computer and Materials Science, but with little participation from the Department of Aeronautics and Astronautics. The center was engaged in high power, multi-million-dollar plasma experiments, mainly funded by the US Department of Energy and with no NASA involvement. They included the ALCATOR-C Tokamak and its eventual follow-on, ALCATOR C-MOD. Both machines were the offspring of MIT's highly successful ALCATOR experiment, a small tokamak originally conceived by MIT Physicist Bruno Coppi in the late 1960s. In the late 1970s, experiments with this device, led by Dr. Ronald Parker and his team, produced the famous "Alcator Scaling," a tokamak fueling approach that led to record plasma densities, a key requirement in the quest for fusion. These results generated a renewed sense of optimism among fusion scientists about the feasibility of the tokamak approach to controlled fusion.

Although plasma rockets were clearly a departure from their mainline activities in magnetic fusion, the group understood the physics of the VASIMR® and saw the plasma propulsion work as a new opportunity to branch out into other areas of high density plasma engineering and research. The MIT-PSFC had also begun experiments on the TARA tandem magnetic mirror, a linear configuration that had also shown promise in two earlier devices: the Tandem Mirror Experiment (TMX) at the Lawrence Livermore National Laboratory in Livermore, California and the Phaedrus Tandem Mirror at the University of Wisconsin in Madison. The Plasma Propulsion experiment was much more modest in scope and objectives than these devices, but had many of the same features of interest to plasma scientists. More importantly, its funding, originating from NASA and the US Air Force and not from the DoE, was not viewed as a threat to the lifelines of the other experiments at the PSFC.

There was sufficient MIT expert support for the validity of the propulsion experiment to overcome the expected dose of skepticism that had surfaced around the construction of the device. Early supporters included distinguished plasma physicists, such as Professor Bruno Coppi and Professor Lawrence Lidsky, who considered the technology challenging but anchored in sound physics and worthy of study. PSFC Director, Professor Ronald Davidson, and ALCATOR-C experiment leader, Professor Ronald Parker, were also early supporters, who found suitable laboratory space for Dr. Chang Díaz and his partner, Dr. T. F. Yang, to set up their experiment. Other early MIT advocates included Dr. Daniel Cohn, an expert in plasma devices who had worked with Dr. Chang Díaz during his early graduate student years, and Dr. Joseph Minnervini and Dr. Joel Schultz, both experts in advanced superconductors. They were both able to foresee the technology advances that could enable the construction of a VASIMR® superconducting magnet strong enough to support the system requirements and light enough for flight. On the theory side, MIT Professor Kim Molvig, a theoretical plasma physicist, became a strong supporter and valuable collaborator in the project for many years, supervising the PhD thesis work of Mr. Warren Kruger, one of Dr. Chang Díaz's graduate students, as he continued to make progress on his study of the dynamics of the plasma exhaust and his numerical simulation of the plasma flow in the magnetic nozzle.

Experiment construction got underway in 1987. Because of budget and vacuum infrastructure limitations and the need to use existing plasma equipment on loan from MIT and DoE, the early device would still be far from a rocket. Instead, the configuration more closely approached the topology of a tandem magnetic mirror, a familiar configuration in fusion that would serve as a "physics test bed." A linear arrangement of current rings formed a solenoid (the central cell), with two smaller but stronger pairs of current rings at each end (the end cells). The resulting architecture produced a sort of "magnetic sausage," with one end more pinched than the other. Plasma could be injected through the stronger "pinched" end, heated to the desired conditions in the central cell and exhausted at the other end. In this way, the physics of plasma generation, heating and acceleration could be characterized before embarking on more ambitious experiments.

The construction of the device, begun in 1987, was more than an exercise in precision engineering; it was an early demonstration of the manufacturing frugality and innovative procurement that have remained key characteristics of the VASIMR® team to this day. In order to provide sufficient vacuum volume to house the plasma discharge, the group developed a rather simple, yet high precision, cylindrical stainless steel vacuum chamber, with appropriate provisions for diagnostic windows and RF feedthroughs. The chamber, built under a small internal contract with the MIT-PSFC Machine Shop, was designed to fit inside the bore of a solenoidal assembly of magnetic coils. The magnetic coils were also designed by the VASIMR® team, but with the pro-bono technical advice of helpful experts from MIT's Francis Bitter National Magnet Laboratory, an extraordinarily valuable input from a world-class organization conveniently located next door to the PSFC. For Dr. Chang Díaz, the establishment of a permanent VASIMR® experimental team at MIT was a high priority objective. The team would lend stability and continuity to the project, as Dr. Chang Díaz had to continuously balance his rocket research time with his astronaut training responsibilities.

The VASIMR® experiment buildup took the better part of a year, with the many details of the system integration and initial tests consuming a great deal of time and attention on the part of Dr. Chang Díaz and his team. The Astronaut Office management, however, recognized the importance of this work. Through careful and efficient schedule management and despite his astronaut training constraints, Dr. Chang Díaz was allowed to travel to Cambridge for several days every month to guide the program and supervise the small in-house team, composed of his colleague, Dr. T. F. Yang and graduate student Mr. Scott Peng. The virtual research capability that Dr. Chang Díaz had established at his Astronaut Office post also facilitated the communication among the researchers.

By the middle of 1988, Dr. Chang Díaz was assigned to start training to fly on his second space mission, STS-34, due to launch in October of 1989. The mission's primary objective was the on-orbit deployment of the Galileo spacecraft, a nuclear-powered mission to the planet Jupiter. As a veteran flyer with eight full years of training under his belt, his preparation for this mission focused on the new flight procedures which had been implemented post-*Challenger*. There were also some important aspects of the Galileo mission that made the flight unique, including a higher inclination orbit than the nominal 28-degrees utilized in most missions, and the special handling of two plutonium radioisotope thermoelectric generators (RTGs) – the first of their kind to fly on board the Shuttle – to power the Galileo on its long journey to Jupiter.

After the *Challenger* accident, the Centaur upper stage that was originally slated to propel Galileo to the Jupiter system had been eliminated from the Shuttle manifest on the grounds of safety. Its replacement was the less powerful, but safer, Inertial Upper Stage (IUS), a proven rocket managed by the US Air Force. This change necessitated charting a new, more complex and circuitous trajectory to Jupiter that would bring Galileo to a prior close encounter with two other planets: Venus and Earth. The close encounters were needed to secure "gravity assists," or gravitational kicks, from these bodies that could supplement the lower propulsive performance of the IUS. In its new trajectory after deployment from the Shuttle, Galileo would first loop around Venus for its initial gravity assist, return to Earth for its second and then loop back to the inner solar system before passing by Earth again for a final kick. The three-kick maneuver was called VEEGA (for Venus, Earth, Earth, Gravity Assist) and would, together with the impulse from the IUS, provide Galileo with the required velocity to reach Jupiter six years later.

Such a complicated cosmic ballet implied a very small orbital window for the deployment of the spacecraft and this was reflected in the intensive training of the Shuttle crew. The deployment of Galileo from the Shuttle payload bay had to occur with clockwork precision, very early in the flight. The crew's training had begun in earnest after the Shuttle's return to flight, once again curtailing Dr. Chang Díaz's availability to support the VASIMR® experiment. Fortunately by this point, the MIT research team was fully consolidated and Dr. Ted Yang was able to provide continuity to the experimental program during Dr. Chang Díaz's mission training. The first VASIMR® experiments were initiated in a small laboratory within the MIT-PSFC in early 1988. The apparatus did not look at all like a rocket. In fact, the massive electromagnet structure powered by thick copper cables carrying thousands of amperes of electric current, together with the complex assembly of pumping ports, vacuum feedthroughs, diagnostics, RF generators and impedance matching networks, capacitor banks and other laboratory paraphernalia, made the device look like a Rube Goldberg machine. Traditional electric propulsion experts, viewing the set up on occasional visits to the MIT laboratory, would often examine it with bemusement and barely concealed chuckles.

The coils consisted of stacks of #1 solid copper conductor pancakes, encased in epoxy resin. They were manufactured, under the watchful eye of Dr. Yang, by a small company by the name of MagneCoil in the town of Peabody, Massachusetts. The assembled stacks were each individually housed in a stainless steel envelope, which also served as a liquid nitrogen plenum to provide cooling. The assembly and final welding of these enclosures was done by yet another small company by the name of Atomic Limited, located near Central Square in Cambridge, who served the needs of many small-budget science experiments at MIT. Each enclosure was designed so that the coil stack inside could be cooled independently, while its current could also potentially be controlled independently. This design provided a great deal of operational flexibility and control in the shaping of the resulting magnetic field.

Two families of coils were manufactured: a set of four small-bore high field stacks, two for each of the end-cells, and a set of eight large-bore, lower field stacks for the central cell solenoid. At the highest point, the magnetic field in the end cells was strong, reaching nearly one Tesla. This required thousands of amperes of electric current with associated amounts of resistive heating in the copper windings. To reduce the thermal load, the coils were chilled with liquid nitrogen. The plasma formed in the device was held by this strong field and kept away from nearby structures. This allowed the process of plasma heating by means of electromagnetic waves to take place.

Looks aside, the MIT device was an excellent and low-cost scientific testbed with which to examine the controlling physics of the VASIMR® engine and begin to prove its viability. It was critical to demonstrate the stability of an axially asymmetric plasma in a linear device, bound by the injection of propellant at one end and the magnetic nozzle on the other. It was also critical to demonstrate that the plasma could be generated and injected efficiently, at the required rate, and that it could be further heated to the desired rocket conditions by the process known as Ion Cyclotron Resonance Heating (ICRH). Already, in the 1980s, ICRH was a well-known heating approach in magnetic confinement fusion devices, one that had gained a great deal of popularity in large tokamaks, where RF antennas inside the toroidal chamber could deliver up to several megawatts of RF power to a high density hydrogen plasma. On the other hand, experience in the use of ICRH in linear machines was limited to a few devices, including the Phaedrus and TMX experiments. Linear devices for fusion were less numerous and continued to lose ground to the more popular toroidal systems, such as tokamaks and stellarators. Moreover, the uniqueness of the VASIMR® configuration, where plasma needs to rapidly flow axially from the injector to the nozzle in order to create the rocket exhaust, added relevance to the experiment.

To provide a suitable dump for this exhaust, the team needed a large enough vacuum chamber as well as high-throughput vacuum pumps. One of two, 3-meter long, 1.5-meter diameter cylindrical aluminum chambers, designed to house each of the end-cells of the TARA fusion experiment, had become available, and was acquired by the team to serve as a plasma dump tank for the rocket. The chamber was to be attached later in the experimental program, when sufficient funding could be obtained to purchase the vacuum pumps needed to maintain the large plasma throughput.

The MIT experiment was modest in scope and had to exist amid a sea of skepticism. The NASA portion of the funding was managed by JPL, but there was no advocacy for the project there. On the contrary, from the very beginning of the MIT program rumors began to surface, amid mostly collegial conversations at technical meetings and conferences, about strong adverse undercurrents that had begun to undermine the project at the JPL and Air Force management levels. Even some in the leadership of MIT's Department of Aeronautics and Astronautics, who were not participating in the experiment at the PSFC, continued to voice their lack of support for the project and lobby NASA and the AFOSR for its cancellation. Interest in the early VASIMR® was also absent at NASA's Glenn Research Center (GRC) in Cleveland, Ohio and The Marshall Space Flight Center (MSFC) in Huntsville, Alabama. The former had laid its bet on more conventional electric rockets, such as ion engines and its newer Russian cousin, the Hall thruster, as well as in reviving NERVA, the defunct 1960s nuclear-thermal rocket. MSFC, on the other hand, was focusing on a much larger program of chemical propulsion and had not expressed much interest in electric propulsion.

On top of the increasingly adverse political undercurrents, the resource demands of even such a simple experiment were significant, so finding the funding to keep the project operating was always a challenge. To reduce costs, the device operated in pulses of a few seconds

3.4 First MIT experiment, circa 1987

in order to ensure adequate thermal control within the modest cooling capabilities of the laboratory. Fortunately, some level of funding and, more importantly, advocacy for the experiment, came from two early supporters at NASA in Washington DC: Dr. Gary Bennett, Manager of the Office of Advanced Programs and Technology, and Mr. Earl Van Landingham, Deputy Chief of Propulsion and Power at the Office of Space Technology. While programmatically fragile, the combined NASA-AFOSR funding for the MIT project topped approximately $300,000 in its best year, an amount sufficient to get the experiment built and sustain a modest level of operation. But headwinds continued to build up. A new director at AFOSR, Dr. Mitat Birkan, had taken over from Dr. Caveny and, somewhat reluctantly, had agreed to provide continuation funding for the VASIMR® project for at least one more year. He had advised Dr. Chang Díaz of the strong opposition to the project that was building up among his main customer base, the established electric propulsion community.

Plasma discharges at up to 10 kW were accomplished in this system, using argon and hydrogen as feedstock gases. The plasma density and temperature were not impressive by fusion standards; however, the discharge was stable and well behaved and provided an excellent baseline upon which to build. It was a start.

REFERENCE

1. *A Supersonic Gas Target for a Bundle Divertor Plasma*, F. R. Chang and J. L. Fisher, *Nuclear Fusion*; v. 22(8) p. 1003-1013; ISSN 0029-5515; Aug 1982

4

Probing the Physics

The first VASIMR® experiments at the Massachusetts Institute of Technology (MIT) had demonstrated a stable, moderate density plasma discharge, generated using small amounts of microwave power. The team also embarked on early attempts at plasma heating using ion cyclotron resonance techniques and conducted momentum measurements of the plasma escaping from the end cell. The latter technique utilized a carefully balanced pendulum force sensor that provided surprisingly clean measurements. That force sensor became a precursor to a highly advanced thrust measurement diagnostic that is in use today.

In the late 1980s, Dr. Chang Díaz continued to develop the VASIMR® concept and write proposals for additional funding. He and Dr. Yang presented the results of their experiments and simulations at major electric propulsion meetings in the US and abroad. However, the "giggle factor" towards the plasma rocket project remained high. Lacking the required skills in plasma physics, it was difficult for the traditional chemical propulsion community at NASA to visualize the promise of the engine. In addition, the massive structure of the MIT experiment, with its large vacuum infrastructure, World War II vintage RF generators and heavy conventional copper magnets, made it difficult to envisage how such a contraption could ever hope to fly.

SEEKING CULTURAL CONVERGENCE

The traditional electric propulsion community also continued to find the VASIMR® engine far from their comfort zone. Perhaps more significantly, the new rocket represented a potential financial drain on the already limited budget of federal dollars available for electric propulsion. There were countless technical arguments brought forward to buttress the non-viability of the concept. While nonetheless damaging, these arguments were generally highly simplistic and lacked substance and depth, being more a testament to a lack of knowledge regarding the physics of the device. For example, the ionization of the propellant was considered to be too energy intensive and the ionization rate inefficient, making fuel consumption too high for good thruster performance. Magnetic nozzles were

© Springer International Publishing Switzerland 2017
F. Chang Díaz, E. Seedhouse, *To Mars and Beyond, Fast!*, Springer Praxis Books,
DOI 10.1007/978-3-319-22918-8_4

also poorly understood and their magnet structure was viewed as too heavy and bulky to ever be flight worthy. It was difficult for these scientists to anticipate the gradual accretion of enabling technologies that would come of age in the next decades to make the engine's design viable. At the same time, because of its genesis in magnetic fusion, the VASIMR® engine was wrongly perceived in some circles as requiring nuclear power, or worse yet as a fusion rocket, a designation born of technological naiveté that, along with other such fanciful devices, automatically exiled it to the safe – and irrelevant – realm of the future.

But the VASIMR® team believed the technology was closer to the present. They understood the boundaries of the engine's physics and recognized that a decade or so of advances in materials, miniaturization, and high temperature superconductivity would bring them to a technological landscape where the system would cease to be "a thing of the future." A major convergence of two plasma physics cultures was needed: electric propulsion and magnetic fusion. It was precisely that convergence that Dr. Chang Díaz and his early team had set out to achieve. There were, however, many more organizational impediments to overcome.

In the 1980s, two NASA field centers maintained active experimental rocket propulsion programs: The Marshall Space Flight Center (MSFC) in Huntsville, Alabama and the Lewis Research Center (now the Glenn Research Center (GRC) at Lewis Field) in Cleveland, Ohio. Another institution closely associated with NASA, the Jet Propulsion Laboratory (JPL) of the California Institute of Technology, also maintained an active program in rocket propulsion. Of those, only GRC and JPL were significantly active in the field of electric propulsion. There were also small activities, sponsored by the US Air Force and NASA, at several universities, such as MIT, Princeton and the University of Michigan, and at private companies such as Aerojet (today, Aerojet-Rocketdyne), Pratt and Whitney and the former TRW. Small research efforts were also ongoing in Europe, China, Japan and the former Soviet Union. Most of the work centered on gridded ion engines and the Magnetoplasma Dynamic (MPD) engine, a DC, high power electromagnetic device. Russia was perfecting a new grid-less ion engine design known as the Hall Effect Thruster (HET). But these efforts were relatively small in comparison with chemical propulsion activities and as a result, electric propulsion research teams in the US competed fiercely for scarce federal funds.

It was against this backdrop of constrained resources that the VASIMR® engine concept made its initial and unwelcome entry into a community with little financial wiggle room. It was no surprise then that the MIT experiment, enjoying little institutional advocacy, gradually became a programmatic orphan. A suitable and protective home was needed to nurture the technology during its fragile embryonic phase, before it became easy prey to the budget ax. Dr. Chang Díaz began to develop a plan to bring the project to the only NASA center that would offer it some degree of shelter; his home base, the Johnson Space Center (JSC) in Houston. There were, however, a number of complications associated with making this move and a bit of historical digression is required in order to add some color to the general picture.

4.1 STS-61C, January 1986: On the flight deck of the Space Shuttle *Columbia*, Dr. Chang Díaz takes some time off to look at the world below on his first flight into space.

FROM TRAGEDY, CHANGE

The *Challenger* tragedy of January 28, 1986 was a sobering event for the nation and the world. For Dr. Chang Díaz and many of his fellow astronauts who lived the euphoria of the early Shuttle years, it was "the day we lost our innocence." The accident occurred just 10 short days after Dr. Chang Díaz and his crew had returned to Earth on the Space Shuttle *Columbia*. As was probably true with all of his fellow astronauts, the personal impact he suffered from the tragedy was profound and lasting. The *Challenger* disaster marked a period of deep introspection into the relationship between the humans and their new space flying machine. Space exploration was, and continues to be, hard and unforgiving, yet despite the tragedy, human curiosity and the quest for new knowledge remains unabated.

The ripples of sadness and loss resonated deep inside the space agency, a sentiment that facilitated the work of the Rodgers Accident Investigation Board in capturing the main elements that caused the explosion and converting them into concrete and valuable recommendations. The adoption of these recommendations implied a major overhaul to NASA and touched not only flight operations, but the entire management and culture of the organization. Dr. James C. Fletcher, a physicist and former NASA Administrator, was brought back to lead the Agency during the recovery and return to flight effort. JSC in Houston, the home base of the astronauts, was also shaken with major management changes, starting at the very top, including Center Director Gerald D. Griffin, who was quickly replaced in January of 1986 by Jesse W. Moore in an interim capacity for a few months, prior to transferring the leadership to Dr. Aaron Cohen, an unassuming but brilliant aerospace engineer and propulsion expert. Dr. Cohen remained in the post until 1992.

More changes continued to take place even after the Space Shuttle resumed operations and these changes had an effect on the VASIMR® engine research. Deeper in the JSC organization, Mr. George W. S. Abbey, Director of Flight Crew Operations, was reassigned to Washington in 1987 and was replaced by Donald Puddy, a mechanical engineer and former flight director. In addition, Chief Astronaut, Capt. John W. Young, was reassigned and succeeded by Capt. Daniel Brandenstein, another Navy pilot and Shuttle commander. Both Mr. Abbey and Capt. Young had been two of the strongest NASA supporters of the VASIMR® project since the beginning, and their moves to new positions brought a new set of boundary conditions to be considered.

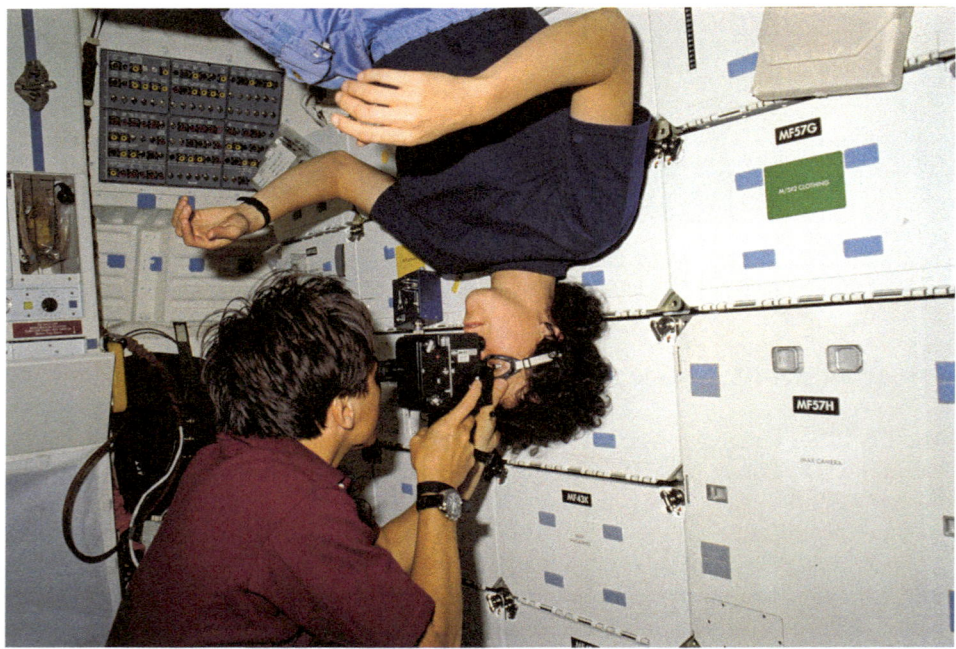

4.2 STS-34, October 1989: On the mid deck of the Space Shuttle *Atlantis*, Dr. Chang Díaz examines the retina of his fellow crewmate, Dr. Ellen Baker. In what probably became one of the earliest examples of telemedicine from space, live images of Dr. Baker's retina were transmitted on the telemetry stream to NASA flight surgeons in Houston Mission Control

Change, however, can also bring opportunity. The seismic jolts in the NASA management deeply shook the old order and prompted a great deal of agency restructuring. The aftershocks also stimulated much needed discussion about NASA and its future and provided opportunities for the VASIMR® team to encourage the new NASA leadership – and some of its key outside advisors – to move past its focus on operations and return to the realm of cutting edge technology, specifically space power and advanced propulsion. A few months before the Space Shuttle returned to flight, Mr. George Abbey was appointed Deputy Associate Administrator for Space Flight at NASA Headquarters and became NASA's representative to the "Synthesis Group," a team headed by former Apollo astronaut, Lt. Gen. Thomas P. Stafford, which was charged with defining strategies for returning to the Moon and landing on Mars.

In July 1991, Mr. Abbey was appointed senior director for civil space policy for the National Space Council in the Executive Office of the President. President George H. W. Bush reestablished the National Space Council by executive order, led by Vice President Dan Quayle. Part of its job was to find a direction for America's space initiatives in a time when the nation would no longer be engaged in a technology race with the Soviet Union. The Council began to see several unique opportunities for engaging the former Soviet Union in a space station program. In 1992, with the arrival of Daniel Goldin at NASA's helm, Mr. Abbey was named Special Assistant to the Administrator, a role that set the stage for his return to JSC; first as Deputy Center Director in 1994 and then as JSC Director in 1996.

During his work in Washington, Mr. Abbey arranged for Dr. Chang Díaz to brief members of the Synthesis Group on the advantages of nuclear-electric propulsion in general and VASIMR® in particular. Also during this time, another JSC supporter, Mr. Jay Greene, had made arrangements for Dr. Chang Díaz to meet and discuss the plasma rocket with Dr. Michael Griffin, another member of the Synthesis Group, who had been working within the Pentagon's Strategic Defense Initiative, also known as President Reagan's "Star Wars" Program. Jay was a former Shuttle flight director who had worked with Dr. Chang Díaz in Houston's Mission Control during the latter's tenure as Capsule Communicator (CAPCOM). Jay was smart and direct and the two had developed a professional friendship, based on mutual respect and admiration. As Ascent Flight Director during the *Challenger* flight, Jay was deeply and directly impacted by the tragedy. He left the flight director track and moved to technical positions in management, as head of the Safety Office and Chief Engineer. It was in that capacity that he summoned Dr. Chang Díaz to a meeting with Dr. Griffin, a young and visionary space engineer and scientist who was visiting JSC as part of his work in the Synthesis Group. Griffin immediately understood the concept of the VASIMR® engine and its future potential. His comment to Dr. Chang Díaz at the end of the briefing summed up his reaction to the project: "Franklin, that's brilliant." Years later, Dr. Griffin would become NASA administrator.

Administrator Fletcher departed and, after a short acting tenure by Mr. Dale Myers, retired Admiral Richard Truly became the first former astronaut to lead the space agency, a post he held from 1989 to 1992. He was succeeded by one of NASA's longest lasting administrators, Mr. Daniel Goldin, who spent nearly a decade at the helm and consolidated the post-*Challenger* agency at the end of the Cold War. He initiated an era of strong international collaboration in human space flight, putting his signature to a multilateral agreement with the major space-faring nations, including the former Soviet Union. These actions opened a path for the first meaningful global teamwork in space, first with the Shuttle-MIR Program, a collaboration that ultimately enabled a number of docking missions between the Space Shuttle and the Russian Space Station MIR, and later with the construction of the far more ambitious International Space Station (ISS).

By late 1989, and amid all of these changes, Dr. Chang Díaz had completed his second space mission and Galileo was well on its way to the Jupiter system. With two flights under his belt, Dr. Chang Díaz was anticipating an extended period of Earth-bound duties within the Astronaut Office, including returning to his research on the VASIMR® engine and supporting other upcoming missions from the ground. However, just a few short months after his return to Earth, he was assigned to his third flight. The assignment was unusually early, as the mission would not fly until 1992; however, the mission merited the allocation of significant additional training for the crew, as its primary objective, the deployment of the first electrodynamic tether in space, was considered highly complex, both technically and operationally. By its nature, the flight involved a very strong plasma physics component, and required a deep knowledge of the intrinsic flight capabilities of the Space Shuttle orbiter. It was deemed by the leadership of NASA's Flight Crew Operations Directorate that Dr. Chang Díaz's skill set was ideally suited for this mission. The mission was called Tethered Satellite System Mission-1 (TSS-1), and was a joint project between NASA and the Italian Space Agency (ASI), as well as other research institutions both in the US and Europe. It was a sign of things to come as Shuttle operations increased in complexity. The orbiter was evolving from a simple, human-tended satellite deployment truck, into an increasingly versatile scientific platform, where astronauts were gradually evolving into the operational scientists that Dr. Chang Díaz had envisioned. Despite the horrendous loss of *Challenger*, and by building on the lessons of that loss, the Shuttle Program had become stronger and was quickly reaching full operational maturity.

The tethered satellite was the brainchild of Professor Giuseppe Colombo, an Italian space scientist from the University of Padua, who proposed that a spacecraft orbiting the Earth in a reasonably low inclination orbit could generate electricity from its motion by deploying a long conducting wire perpendicularly, up or down to its direction of motion. As it moved along its orbital path, the conductor would cut through the Earth's magnetic field, creating a sort of "cosmic dynamo" that could produce electricity at the expense of its motion. The reverse process, namely a motor, was of course also possible in theory. With a suitable electrical power source, such as a solar array with the right polarity, a current could be made to flow in the wire, such that a propulsive force could be imparted to it, thereby generating electric propulsion… without fuel! This, of course, was an extraordinary proposition and while it only worked near planets with magnetic fields, its development could be of considerable use in orbital operations.

For an appreciable effect, the wire had to be rather long – 20 km was the chosen length for the mission – and carefully deployed to avoid entanglement with the ship. The ionized medium in which the Shuttle flew is an electrical conductor, which provided exactly the needed environment for electricity to flow. However, to allow the current to actually flow as the system moved, the ends of the wire had to be equipped with good electrical "brushes" to make electrical contact with the local "ground," namely, the ionosphere. The end contacts were rather unusual: The Space Shuttle itself, at the lower end, and a 50kg spherical metallic satellite at the upper end.

4.3 STS-46, August 1992: Dr. Chang Díaz (upper right) and his fellow crewmates; Commander Loren Shriver (upper left), Pilot Andy Allen (top center), Mission Specialists Claude Nicollier (far left), Jeffrey Hoffman (second left) and Marsha Ivins, and Italian Payload Specialist Franco Malerba aboard *Atlantis*. The first flight of the Tethered Satellite System (TSS) suffered from a mechanical jam that prevented the tether from reaching full deployment. The crew also successfully deployed the European Retrievable Carrier (EURECA), a free-flyer experimental platform to be retrieved in a subsequent mission.

It was an experiment of ingenious design, electrically not unlike the configuration of a trolley bus riding on steel rails, or a bump car moving on the metal floor in a country fair ride. The cargo bay of the Shuttle would be equipped with a number of related experiments housed in an experiment pallet. These were designed to examine the charging characteristics of the vehicle itself and its ability to shed electrical charges to the ionosphere. The electricity was expected to flow either passively through the metallic engine bells of the Shuttle main engines, or actively by means of electron guns and plasma generators housed on the experiment pallet.

For a plasma physicist astronaut like Dr. Chang Díaz, preparing for the TSS-1 Mission was a dream assignment that brought together all the elements of his scientific and operations training. But it also provided an excellent opportunity to promote the culture convergence that he was pursuing in the field of electric propulsion. The mission science team included outstanding scientists from nearly a dozen institutions, both in the US and Europe, and many of these plasma and ionospheric physicists had also delved into electric propulsion in their professional careers and were familiar with many of the key players in the EP community.

One of these was Bryan Gilchrist, a Stanford University electrical engineering PhD candidate, finishing his doctoral program under the guidance of Professor Roger Williamson. Bryan became very familiar with the physics of the VASIMR® engine and, upon completing his PhD, took on a faculty position at the University of Michigan, working closely with the electric propulsion team there headed by Dr. Alec Gallimore. Years later, two of the Michigan graduate students, Christopher Deline and Christopher Davis from the Gallimore/Gilchrist team, completed their PhD theses on specific topics of the VASIMR® engine, under Dr. Chang Díaz's supervision. Their work was conducted at what became NASA's Advanced Space Propulsion Laboratory (ASPL), a facility at the JSC in Houston born from the survival instinct of an idea and the vision of a handful of NASA managers.

A NEW VASIMR® HOME IN TEXAS

Active efforts for the creation of the ASPL at NASA's JSC can be traced back to 1989. These initial overtures by Dr. Chang Díaz to relocate the MIT experiment to Houston met with significant institutional disapproval. JSC was certainly not deemed to be a credible propulsion center, either within or outside of NASA, and there were strong sensitivities within the space agency at seeing one field center usurp the mission charter of another. To make matters worse, a propulsion program run by an "astronaut-in-training" and nested within the organizational structure of JSC's Flight Crew Operations Directorate was viewed as going "against the institutional grain" and not considered programmatically viable. In an ominous telephone conversation with Mr. Earl Van Landingham at NASA Headquarters in Washington DC, in which Dr. Chang Díaz requested permission to move the laboratory to JSC, Mr. Van Landingham, recognizing these strong sensitivities, was compelled to tell him: "Under no circumstances will you bring your project to JSC."

More than friendly advice from a supporter, it was a clear directive from the leadership in Washington. Yet Dr. Chang Díaz persisted and by 1993, there were enough "rogues" within the JSC management that the transition from MIT to JSC eventually took place. It was, however, a bumpy transition. In late 1992, JSC Center Director Aaron Cohen, in a bold maneuver to circumvent institutional rigidity, authorized the transfer of the experiment from MIT to Houston. However, not wishing to violate a Washington directive, he proposed to house the project close to JSC, at the University of Houston. He had become convinced of the disruptive nature of the technology but also recognized its potential value to NASA. Earlier that year, he had assigned his Chief Technologist, Dr. Kumar Krishen, to work closely with Dr. Chang Díaz to accomplish the equipment transfer as quickly as possible. After several high level meetings with university officials, the school agreed to entertain housing the project.

It was a race against time, as the need for the move had suddenly become an emergency. In early 1993, the VASIMR® detractors had gained ground in convincing the NASA management in Washington that the MIT-PSFC experiment should be terminated. Some of these detractors had significant clout within the space community. Therefore, within a couple of weeks, the wheels had been set in motion and a hastily assembled Project Review Panel of "experts" was convened at JPL to advise NASA and the Air Force on the worthiness of the project. Dr. Chang Díaz and Dr. Yang were quickly summoned to Pasadena, California, with no preparation time, to defend the program in a closed-door review.

As expected, the review panel quickly recommended the termination of the MIT project to NASA and the AFOSR. It was clearly a frontal attack by a hidden enemy and a major setback. The VASIMR® team was not allowed to review the findings of the panel, nor to provide a rebuttal. All hope for an appeal to the ruling was lost. The relocation to Houston had suddenly become a matter of survival. In a telephone call from Cambridge, Acting PSFC Director, Professor Dieter Sigmar, informed Dr. Chang Díaz that the MIT plasma propulsion project had been terminated by the JPL management and would be disassembled and its equipment allocated to other programs. However, it was also in that telephone exchange that Dr. Chang Díaz implored Professor Sigmar to keep the experiment hardware intact and off limits for just a couple more days. Three NASA JSC trucks were already en route to Boston to retrieve the equipment for transportation to Houston for reassembly at a new laboratory supported by the University of Houston. Professor Sigmar did his best to comply with the request.

Homeless disassembled experimental equipment does not stay in one place very long, as others are quick to lay claim to the idle components. It was common practice that inactive equipment under the care of the PSFC, but belonging to the US Government, could be shuffled around in support of other government projects. Indeed, it was through this unwritten but accepted protocol that the refurbished end-cell from the TARA experiment became the plasma rocket's first vacuum chamber. In the summer of 1993, the news of the terminated plasma rocket experiment traveled fast, unleashing a volley of claims from other experimental teams for choice parts of the inactive apparatus. Timing was of the essence. It was imperative, if there was to be any hope of reassembling the experiment, that the hardware could be kept together long enough to transport it to Houston.

Fortunately, the team succeeded in keeping most of the hardware intact for long enough. The MIT experiment was disassembled and crated over an intense weekend of non-stop work by Dr. Yang and several graduate students, with the aid of MIT-PSFC's Chief Technician, Mr. Frank Silva and some volunteers from his team. An extraordinarily helpful supporter of the project, Mr. Silva understood quite well the predatory side of experimental teams within a research facility when it came to prime dismantled equipment sitting around for the taking. In that fateful weekend, several costly items disappeared overnight to scavengers from other projects; nonetheless, most of the hardware survived. Thus, on a rainy Monday in the summer of 1993, three large tractor-trailers departed Boston bound for Houston with more than 30 tons of experimental hardware and the hope for a new beginning for the project.

The plan for finding a new home was still in development as the trucks headed south. It involved the creation of a new Advanced Space Propulsion Laboratory (ASPL), led by Dr. Chang Díaz, within the infrastructure of the University of Houston's Advanced Superconductivity Center, led by Dr. Paul Chu. Unfortunately, Dr. Chu had left on an extended sabbatical at the Chinese University of Hong Kong and the remaining Center management did not show much interest in the propulsion project. A ray of hope appeared when Dr. Alex Ignatiev, a colleague of Dr. Chu and head of the university's Space Vacuum Epitaxy Center (SVEC), became interested and began to seek ways to incorporate the new ASPL within his organization. His efforts bore some fruit, as SVEC soon identified an off-campus site for the experiment. Unfortunately, inspection of the space quickly showed that it was insufficient to meet the laboratory's requirements.

The developing relationship between the plasma rocket project and SVEC was extremely tenuous and was barely being held together by Dr. Ignatiev with the assistance of Dr. Chang Díaz and Dr. Krishen. The latter's mandate to help establish the ASPL had been weakened by the impending departure of Center Director Cohen. Nonetheless, the exploratory meetings continued. The scientific affinity between Dr. Chang Díaz and Dr. Ignatiev was strengthened as Dr. Chang Díaz was named Payload Commander on STS-60, his fourth space mission. The mission, to be flown on the Space Shuttle *Discovery*, was the first of the Shuttle-MIR program and included Sergei Krikalev, an experienced Russian

4.4 STS-60, February 1994: On the flight deck of the Space Shuttle *Discovery*, Payload Commander, Dr. Chang Díaz, reviews and organizes the early morning tele-printer messages from Houston's Mission Control. In an age before electronic tablets, written schedules and procedures came to the crew via fax and tele-printer paper, difficult to manage in zero-g. STS-60, commanded by future NASA Administrator Charles F. Bolden Jr., also marked the beginning of a stable US-Russia space relationship, leading to collaborative MIR and ISS programs.

cosmonaut and the first Russian flyer to serve as a member of a Shuttle crew. The mission commander was Charles F. Bolden Jr. It would be the second space mission Dr. Chang Díaz and his close friend Commander Bolden would fly together. Many years later, Charlie Bolden, by then a retired Major General of the US Marine Corps, would be appointed by President Barack Obama as the 12th Administrator of NASA.

The STS-60 flight included a fairly extensive battery of scientific investigations in Life Sciences and Microgravity, most of which were housed in the SpaceHab module, one of the first initiatives in commercializing space. Built and operated by SpaceHab Industries, the facility was a pressurized extension to the Shuttle cabin. It was modular in design and could be further extended to occupy the entire payload bay of the Shuttle if required. In this mission, however, the second one flown, the module was a small unit, fitted in the forward quarter of the Shuttle payload bay. A pressurized tunnel connected the SpaceHab with the Shuttle cabin and allowed the crew easy on-orbit access to the mini laboratory.

An important payload on STS-60 was the University of Houston's Wake Shield Facility (WSF). The experiment came from Dr. Ignatiev's organization, SVEC, and featured a 3-meter diameter "flying saucer-like" stainless steel disk that was to be deployed from *Discovery* using the Canadian robotic arm. The WSF was a robotic demonstration of semiconductor manufacturing under ultra-high vacuum, where specialized composite materials could be built, one atom layer at a time, without the fear of environmental contamination. These techniques, pioneered by Dr. Ignatiev and his team at SVEC, were expected to bring a revolution in electronics and other applications of specialized materials. By flying face on through the ionosphere at the 7.5 km/sec orbital speed, the WSF would "plow" through the Earth ionosphere, creating a "shadow" of ultra-high vacuum behind it, thus creating the clean environment needed for the manufacturing to proceed.

The crew of *Discovery* also included Dr. Ronald Sega, a rookie astronaut who had been selected by NASA in 1990 while serving as Associate Professor of Physics at the University of Houston. He had become affiliated with SVEC as Co-Principal Investigator on the WSF program and, together with Dr. Ignatiev, helped cement a strong research link between the crew and the university. As he continued to search for a home for the VASIMR® project, Dr. Chang Díaz hoped the SVEC relationship could help motivate interest from the university in the plasma rocket.

Sadly, despite numerous additional meetings, some with hopeful signs, collaboration and institutional support from the university's top management never materialized. It became quite clear that the university was not prepared to embark on such an ambitious new line of research. Additionally, there was no budget available to operate such a large-scale facility. The university's experience in experimental plasma physics was limited to modest activities in ionospheric research on sounding rockets and balloons and had not delved into high-density magnetized plasma propulsion. Discussions for a similar laboratory at Rice University were also disappointing. Upon arrival of the three tractor-trailers, bearing the totality of the equipment from Boston, it was evident that the hoped-for University of Houston home for the project was not to be.

The University took no further action and with no home for the equipment, Dr. Chang Díaz was forced to put the hardware in a temporary storage space at the Johnson Space Center (JSC) allocated by Center Director Cohen. The equipment remained in storage for nearly two years as more management changes continued to happen at JSC. In August of

1993, Center Director Cohen stepped down, and was succeeded by Dr. Carolyn Huntoon, a life scientist who had led the medical space research activities in Houston. Her tenure was also brief, as she transitioned to a leadership position at the Texas Medical Center in Houston, opening the way for Mr. Abbey's return to the top JSC post in Houston.

HOME AT LAST – SORT OF...

The VASIMR® experimental hardware gathered dust for more than a year in the JSC warehouse, until 1994. In that year, the former McDonnell Douglas Aircraft Corporation (MD) lost its bid to build NASA's Space Station Freedom (SSF) and a large, brand new building, built to house the station components prior to shipment to the Kennedy Space Center (KSC), lay empty and with no mission. The building, called the Clear Lake Development Facility (CLDF), was located on the eastern side of Ellington Field, a former US Air Force base turned municipal airport, about 10 miles from JSC and the home of NASA's astronaut flight training operations. JSC's new Center Director, Mr. George W. S. Abbey was making arrangements for NASA to purchase the building from MD to build the new astronaut Neutral Buoyancy Laboratory (NBL), a world-class underwater facility that would be needed to train astronauts on the complex space walks required for the on-orbit assembly of the SSF. The fate of the station itself was bleak, facing certain congressional cancellation. However, in the wake of the break-up of the Soviet Union, a new program, promoted by NASA's new Administrator Daniel S. Goldin, was taking shape. The program involved the morphing of the moribund Space Station Freedom into an ambitious new undertaking to build ISS, a multinational program with a partnership of space-faring nations, including the US, Russia, Japan, Canada, and several European countries.

Within this new framework of international collaboration, during a casual lunch at the JSC cafeteria with fellow astronaut William "Shep" Shepherd in early 1994, Dr. Chang Díaz discussed the "limbo" of his plasma rocket, presently in storage at JSC and with no place to go. A US Navy SEAL and quite familiar with propulsion applications in nuclear submarines, Shep was also a good friend and a fan of VASIMR® as the propulsion system that could enable humans to travel in deep space at high speed. During this informal encounter, Shep, who in 1999 became the first Commander of the ISS, recommended that Dr. Chang Díaz take a look at the CLDF. The gargantuan facility was not being used and would have ample space to set up his rocket laboratory. In addition, the builders had accommodated an uninterruptible electrical power system (UPS), which could potentially be adapted to provide the power requirements of the plasma experiment.

It was a breath of fresh air. That same afternoon, Dr. Chang Díaz toured the empty facility accompanied by MD Facility Managers, Mr. Jim Christian and Mr. Roy Gunn. The building was enormous and mostly unoccupied. A large high bay had been designed to support the pre-shipment assembly of the SSF modules. Adjacent to the high bay were supporting rooms, also quite large, designed to provide facilities for electronics assembly and testing, manufacturing and other purposes. One of these adjacent facilities was a large 100 ft. by 50 ft. clean room, a high-bay space complete with built-in utility compressed air, chemical venting hood, a full size access door and an integrated 5-ton crane that operated along the full length of the space. The full access door was big enough to allow easy access to the three large Continental Electronics FRT-85 RF generators, high power rectifiers and vacuum chamber that were part of the experiment hardware. Despite the fact that it lacked substantial high power capability, the room was exactly what the team needed and Dr. Chang Díaz wasted no time in letting Mr. Christian and Mr. Gunn know he was keenly interested in the space to set up his plasma laboratory. He agreed to initiate the process of obtaining the necessary permissions to move the experimental hardware into the facility.

Fortunately, the starting point for this process was none other than the JSC Center Director, Mr. George Abbey, a long-time supporter of the project. Mr. Abbey had continued with his efforts to arrange for the purchase of the facility by NASA. The assembly of the ISS in space would require an enormous amount of work by spacewalking astronauts and NASA needed to be ready to meet this requirement. The NBL would be the largest indoor swimming pool in the world, where astronauts would train underwater on the complex extravehicular tasks that would be required to carry out the ISS assembly. The government purchasing process, however, would be lengthy, but while the building still belonged to McDonnell Douglas, Mr. Abbey obtained the approval from the company's management and cleared Dr. Chang Díaz to initiate the movement of the equipment and to begin reassembling the rocket experiment. The reassembly, however, was a daunting task for a single person and since the project was no longer under the MIT structure, Dr. Ted Yang could no longer be involved. The MIT graduate students had since completed their theses and were also gone. It was a new beginning and Dr. Chang Díaz needed help.

That help came in late 1994, when Dr. Jared P. Squire, a young MIT plasma physicist and expert in RF plasma heating, contacted Dr. Chang Díaz to inquire about an opportunity to work on the plasma rocket. Dr. Squire had completed his PhD in experimental physics at MIT and had done his research on microwave plasma heating, under Professor Miklos Porkolab, on MIT's Versator Tokamak, a small research experiment in operation at the PSFC. After completing his PhD, Dr. Squire had joined the RF team of the D-IIID Tokamak project, one of the largest fusion experiments in the world, operating at the General Atomics Company in San Diego, California. But Dr. Squire was keenly interested in space propulsion and, as a graduate student, had followed the work on the plasma engine at MIT. He was an accomplished experimentalist, exactly the kind of person Dr. Chang Díaz needed to kick start the project. The only problem was funding.

In a meeting with Mr. Abbey, Dr. Chang Díaz received approval for $50,000 of Center Director Discretionary Funds to begin re-assembling the experiment at the CLDF and to initiate the transfer of the equipment from storage to the new laboratory space. Dr. Chang Díaz also began to conceptualize a mechanism for Dr. Squire's potential employment in the new laboratory. As a civil servant, Dr. Chang Díaz could not directly hire Dr. Squire, but a contractual agreement for technical services through a private company was possible. Such an arrangement could be made through Dr. Yang's one-man private consulting firm, Yang Technologies Inc., of Wayland, Massachusetts.

In December of 1994, Dr. Squire visited JSC and met Dr. Chang Díaz for the first time. They discussed the potential hiring arrangement for Dr. Squire, toured the new, still empty, laboratory space and inspected the hardware in storage at JSC. Dr. Squire visited again in January of 1995 with his wife Lisa, a registered nurse who clearly had to vet a potential family move from San Diego to Houston. By this time, the reassembly process had now begun, as Dr. Chang Díaz had completed the hardware move from storage to the CLDF with the assistance of Mr. Art Rabeau, a graduate student from the University of Houston. The laboratory was named the Advanced Space Propulsion Laboratory (ASPL), with Dr. Chang Díaz as its Director.

The facility, however, was born as an organizational orphan, with no programmatic home or stable funding. On a number of occasions, Dr. Chang Díaz spotted awkward references to his plasma laboratory in the organizational structure of the Astronaut Office; undoubtedly a valiant effort on the part of his direct management to recognize its existence and provide it some level of legitimacy. Any funds made available for the project were managed by the business office of the Flight Crew Operations Directorate. Several individuals among their team of accountants and business managers, most notably Ms. Camille Goodwin, whose continuous attention – spotting targets of opportunity for funding for the facility during virtually the entire life of the project at NASA – was critical to its survival. Later on, Ms. Martha Bishop, also in the Business Office, provided invaluable support, ensuring that any funds that became available were efficiently and expeditiously utilized to keep the project going.

During Dr. Squire's second visit to JSC, Dr. Chang Díaz verbally offered him a position as a research assistant in the new ASPL, but there was a bit of a catch that made Dr. Squire think hard about the proposition. The $50,000 from Center Director Discretionary Funds was a one-time NASA grant to give the project time to seek more stable funding. Dr. Chang Díaz had to explain that, though he was optimistic about securing continued funding, there was no assurance that the present support would be renewed beyond nine months of employment. An offer letter from Yang Technologies Inc., to Dr. Squire was formally sent on February 24 of 1995.

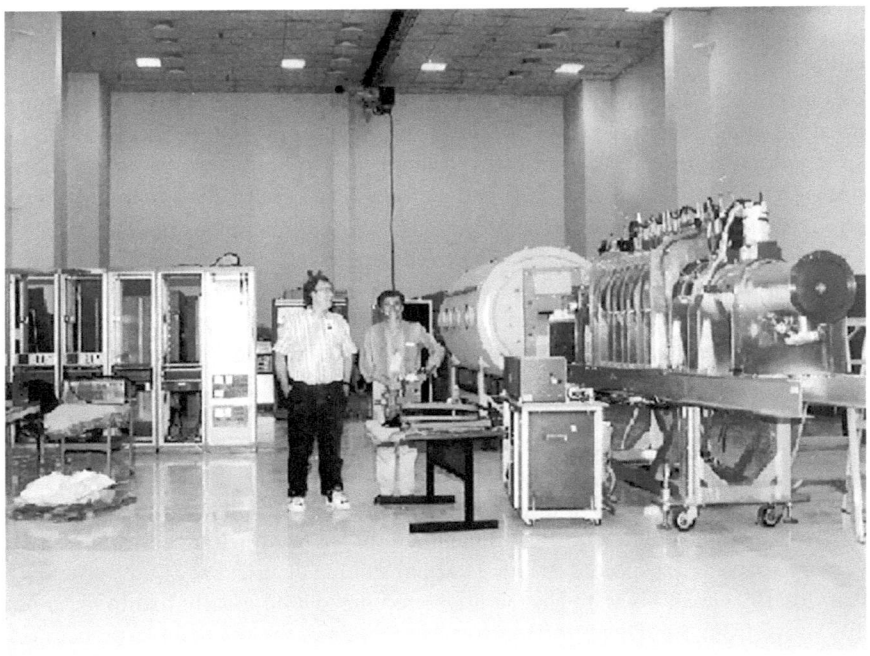

4.5 Dr. Squire (left) and Dr. Chang Díaz in the new NASA-JSC ASPL facility as they began the reassembly process of the plasma rocket experiment (Circa, April 1995).

It was a tough decision for the Squires, but funding uncertainties notwithstanding, Dr. Squire had fallen in love with the project and, after consulting with Lisa, decided to accept the challenge. They arrived in Houston in April of 1995. Dr. Squire had been hired as an employee of Yang Technologies Inc., specifically to support the NASA project at JSC. The company became a contractor to NASA JSC, with Mr. Tomas Krenek as the Contracting Officer (CO) and Dr. Chang Díaz as the Contracting Officer Technical Representative (COTR). The complicated legal arrangement was the only one that could work and be in compliance within the Federal Acquisitions Regulations (FAR). To support his new management role, along with his astronaut training, Dr. Chang Díaz also underwent COTR training.

The new laboratory space was nearly perfect and the location extremely convenient. The facility was located at the end of Space Center Boulevard, a brand new road that has since become a major thoroughfare, linking the Johnson Space Center to points north. For Dr. Chang Díaz, it was a short, 20-minute commute from his office on the main JSC campus to his new laboratory. The brand new facility, however, lacked adequate power, an important requirement for the contemplated experiments. That problem was solved by swift action from Center Director Abbey. He assigned Mr. Daniel Tam, then head of the Space Station Program Office at the JSC, to find a solution to the power issue. As the Clear Lake Facility was being readied to house NASA's NBL, the building clearly needed major modifications anyway and power upgrades were part of the lot. Dan Tam was an agile manager and quickly found the resources to provide the power infrastructure required by the new laboratory. A new underground high voltage line was installed, with provisions for future expansion. The new power line terminated at a high voltage switching module, which was added to the space. Dr. Chang Díaz then issued a small contract to Macroamp, a small, high voltage equipment company in northern California, to manufacture the required transformers, power quality and filtering units that enabled the power to be used within the laboratory. When the installation was completed, the ASPL had an impressive 1.5 MW of installed power capability, sufficient to run the ambitious experiments being contemplated by the VASIMR® team.

Three individuals, Dr. Squire, Dr. Chang Díaz and UH graduate student Art Rabeau, set out to reconstruct the VASIMR® experiment, working from laboratory notebooks and hundreds of photographs of the original experiment brought back from Boston. For Dr. Chang Díaz, these were times of intense activity, as he was also preparing for STS-75, his fifth space mission, scheduled for February of 1996. But the crew training brought an added bonus to the VASIMR® team, as Italian astronaut and STS-75 crewmember, Dr. Umberto Guidoni, a plasma physicist himself, volunteered to help with the experiment reassembly and dedicated many hours of his personal time to helping out in the laboratory. In addition, Dr. Scott Horowitz, the mission pilot and an accomplished mechanic, designed and built a steel frame in his home shop to support one of the microwave transmitters used in the plasma generation.

It was a painstaking and slow effort to reconstruct the intricate vacuum system, magnet power supplies, neutral gas injectors and the microwave power system that generated the plasma in the early experiments. Nonetheless, it was also an exciting time. The experiment was being rebuilt from the ground up, with a much greater capacity and in a far larger and more modern laboratory. The large vacuum chamber, which at MIT was never connected to the apparatus due to lack of funds, was thoroughly cleaned and sand-blasted to eliminate the large carbon deposits that had accumulated on its inner surfaces from its earlier life as part of the TARA experiment.

4.6 Important visitor: Senator (and former astronaut) John Glenn (second from left) visited the early laboratory on May 31, 1995. Other visitors would soon follow, as the facility became part of the VIP tour schedule at NASA JSC.

There was excitement in the air, mixed with some level of fear, as the project began to come to the attention of the JSC community. Many at JSC were in awe of the new technology that had arrived at their center, considering it the next giant leap in space exploration. The new laboratory soon became part of the JSC VIP tour schedule and many dignitaries stopped by to see the new cutting-edge research going on at the center. Many joked that the lights of Houston dimmed when the "mad scientists" working at the CLDF fired the plasma rocket. Others continued to be convinced that the VASIMR® engine was actually nuclear powered, a fusion rocket with its million-degree plasma.

But there were also many who understood the physics of the engine and the significance of the work going on at the ASPL. In that early period, as well as throughout the years at JSC, many NASA JSC employees found ways to help the team through their particular jobs within the Center's organization. This attitude was especially true at the Flight Crew Operations Business Office, where the management always made sure that the ASPL's designated funds were allocated with no delays. The VASIMR® engine project became very popular among the young US Navy ensigns who were deployed to the Astronaut Office as part of their summer experience. Many of these young officers hoped to become astronauts some day and saw the work as a choice assignment. The same was true of other members of the military branches who spent tours of duty at NASA.

An important contribution to the laboratory came from one of these individuals, a US Army officer assigned to support the Vehicle Integration and Test Team (VITT) at JSC.

The VITT supported the Astronaut Office in the final preparations of the Space Shuttle and its payloads prior to flight. As Payload Commander of the second mission of the Tethered Satellite System, his fifth flight into space, Dr. Chang Díaz worked closely with the VITT on the integration of the payload hardware on the Space Shuttle *Columbia*. During periods of free time across several of their joint visits to the Marshall and Kennedy Space Centers, the officer engaged Dr. Chang Díaz in conversations about the plasma rocket. He had developed a keen interest in understanding the particulars of the engine and the needs of the new ASPL. It was during one of these conversations that the officer made Dr. Chang Díaz aware of a small surplus 2.45 GHz microwave transmitter and associated waveguides owned by the US Army. The equipment lay idle and with no identified user at one of the laboratories of the US Army Academy at West Point. The MIT transferred equipment included one such transmitter, but its condition was unreliable. The equipment was important for plasma initiation and the Army hardware would provide an excellent alternative. Within days, the officer had worked through his chain of command to obtain approval for the transfer of the hardware to JSC. The hardware was quickly incorporated and enabled the first plasma shots to be fired in the ASPL facility in August of 1995.

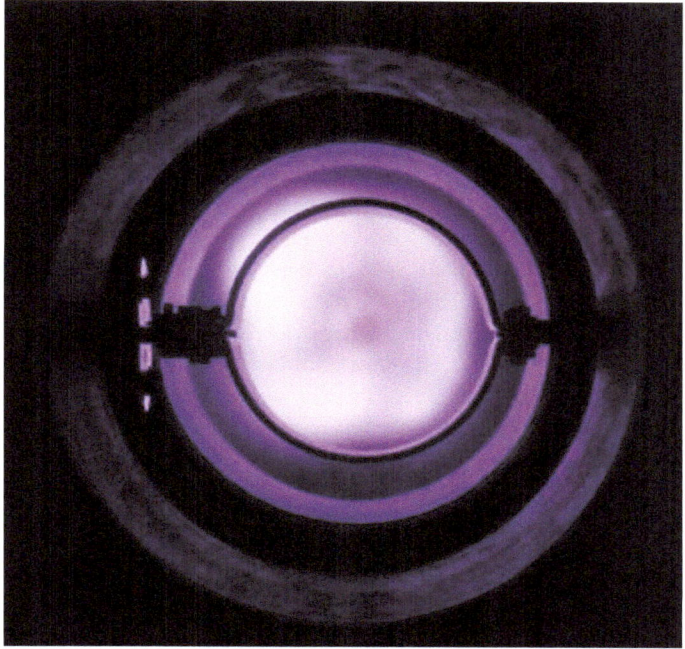

4.7 First plasma at the ASPL was obtained on August 21, 1995.

EXPLORING VASIMR® TRAJECTORIES TO MARS

The configuration of the ASPL made it much more flexible and capable than the previous MIT laboratory. The team added a great deal more experimental flexibility on the plasma production. A second, more powerful microwave transmitter, operating at 14.7 GHz was being incorporated to the experiment at MIT when the emergency move to Houston occurred. This equipment had never reached full integration but, having found its way to Houston, now it could. In addition, a small Magnetoplasma Dynamic (MPD) source was obtained from Dr. Robert Jahn's laboratory at Princeton, which Dr Chang Díaz was considering using as a plasma injector for the VASIMR® engine. The facility was now run by its new director, Dr. Edgar Choueiri, and Dr. Chang Díaz had obtained the hardware in exchange for releasing a small test ion engine from NASA, which Dr. Choueiri required for his experiments.

Progress was also being made on the theory and computational front. Dr. Chang Díaz's approach to space propulsion, using the VASIMR® engine, involved a novel technique which he called "Constant Power Throttling." This method took advantage of the engine's variable thrust and specific impulse to optimize propulsive efficiency throughout the flight trajectory, taking into consideration the gravity field. The solution of this problem in orbital mechanics involved numerical optimization techniques, requiring significant computational power.

To tackle this optimization problem, Dr. Chang Díaz enlisted the collaboration of Ivan Johnson and Ellen Braden, two experienced NASA scientists working in the trajectory analysis group at JSC. Both Ivan and Ellen had developed an interest in the VASIMR® engine concept and wanted to explore its advantages over the traditional chemical and nuclear-thermal propulsion approaches to Mars. In addition to these investigators, Dr. Chang Díaz also enlisted the help of Michael Hsu, one of the Midshipmen from the US Naval Academy at Annapolis who had recently been assigned as a summer intern to the Astronaut Office. Michael compiled the required engine performance assumptions and mission parameters from Dr. Chang Díaz and carried out the computer simulation runs with Ivan and Ellen's assistance. In this way, a small and informal mission trajectory team, led by Dr. Chang Díaz and focused on studying deep space missions with variable specific impulse, was established at JSC. The new team also included Dr. Ted Yang at MIT, who continued to be involved in the project in an advisory role from Boston.

In 1994, the informal trajectory group embarked on an extensive study of mission scenarios for fast human missions to Mars, with VASIMR® nuclear-electric propulsion. The study proceeded from an assumed mission architecture known as "split-sprint." The scheme, variants of which are widely used today in Mars exploration architectures, involves two separate vehicles – one robotic, the other with a human crew – flying roughly two years apart. The robotic vehicle is a relatively slow but high payload cargo ship that pre-positions the human support infrastructure in orbit around Mars and on the planet's surface. The second vehicle is a moderate payload "fast boat" that transports the crew in a much shorter transit time. The roughly two-year interval relates to the two-year cycle over which the two planets reach favorable alignment for the transit, as they circle the Sun in their orbits.

The early VASIMR® split-sprint mission involved a slow robotic tug-boat making the trip to Mars in 180 days with 67 percent payload, followed by a 100-day fast-boat with a

2 percent payload. Both vehicles would use nuclear-electric power with VASIMR® propulsion. The 100-day (101 outbound and 104 inbound) transits allowed for the return to Earth in the same Martian year and a surface stay on the planet of 30 days. Returning to Earth in the same Martian year was considered desirable, in order to free the crew from having to "winter down" on Mars for two years while waiting for the next planetary alignment. The study found that shortening the human transit times to 90 days was also possible, if one sacrificed the same-year return requirement and accepted the two-year Mars "winterdown." This would be a possibility for later on, when a robust and sustainable Martian infrastructure would be in place. The work involved a great deal of iterative computer runs, seeking optimal vehicle configurations and departure windows. Some of the work was also dedicated to the study of contingency scenarios caused by hypothetical failures, which explored the robustness of the VASIMR® engine to critical failures affecting the propulsion system. The results of this study were summarized in a 1995 NASA report, entitled "*Rapid Mars Transits with Exhaust Modulated Plasma Propulsion* [1]," a seminal paper which first described the operational advantages of the VASIMR® engine in the context of a human mission architecture.

The informal group of trajectory analysts was only one example of unofficial institutional support for the new plasma rocket project. Another relationship was established with the Safety Reliability and Quality Assurance (SRQA) Division of JSC. The division was led by Air Force Colonel (Ret.) John Casper, a fellow astronaut who believed that it was important to NASA that the plasma rocket project be supported. He deployed one of his engineers, Mr. D. Scott Winter, to support the ASPL facility. Scott became a part of the ASPL team from the early years and stayed on until the laboratory became a private company in 2005. During his tenure with the project, he supported many of the required certifications needed to maintain a safe operation. Other NASA JSC engineers also assisted the project, such as Andrew Petro, an aerospace engineer from the Office of Exploration, who contributed in many different areas of engineering and analysis and led some of the ASPL's early incursions into superconductivity. There were others from the thermal analysis division of the Engineering Directorate, as well as Dr. John Shebalin, a theoretical plasma physicist who became involved later and provided valuable contributions to the project.

While there were many at JSC who openly supported the new rocket laboratory, there were others who altogether opposed the facility. Some, arguing on programmatic grounds, considered the facility to be a distraction, not belonging to the "mission" of NASA JSC as an "operational" center. Others actively sought to discredit the project as not having sufficient JSC institutional oversight and lobbied to stop the research on the grounds of safety concerns, with the "mad scientists" working on a "dangerous nuclear device." These fears ebbed and flowed over a whole decade of ASPL operations. On one occasion, a safety investigation was conducted on alleged – and unfounded – fears of the ASPL being a cancer node, threatening the health of its personnel and that of the neighboring organizations housed within the CLDF. These local institutional tremors, however, were relatively mild in comparison to the tectonic stresses building up within the electric propulsion community at large, as the VASIMR® team continued to grow and make steady progress on the experimental front. The group had begun to produce numerous publications and conference presentations, which captured the attention of outside organizations working on plasma research, including universities, national laboratories and other NASA centers.

PLASMA WITH ROOM TO GROW

With Dr. Squire's arrival in April of 1995, reassembly of the experiment proceeded very quickly and just four months later, on August 21, 1995, the team achieved the first plasma in the new facility. The ASPL was now operational and producing a stable plasma. Less than one year had passed from that fateful lunch at the JSC cafeteria where Bill Shepherd had made Dr. Chang Díaz aware of the potential new home for his plasma rocket. The experiment was able to produce only a basic plasma discharge, utilizing a small 1 kW, 2.45 GHz microwave generator, but the sight of the bright blue argon plasma through the vacuum window was truly impressive. It was a major success and a psychological boost to the small team, especially given the extremely low budget the project had. Two other individuals had appeared on the scene shortly after the apparatus was reassembled and quickly became part of the group: Mr. Garland "Buddy" Goebel, a technician who worked at the CLDF, and Dr. Edgar Bering, a University of Houston physics professor and student advisor to Art Rabeau. Dr. Bering specialized in ionospheric physics, but also had an interest in electric propulsion. Buddy designed and manufactured unique structural and mechanical interfaces, building from scratch some of the components that had not made the transit from Boston. The meager budget made the team very adept at improvising and finding innovative low-cost solutions to problems. Buddy was also a master at "scrounging," often finding a fresh supply of materials at the local dumpster where they had been discarded by other projects. All of the ASPL furniture, experiment racks, overhead cable trays and some structural materials came from hardware obtained, at no cost, at JSC's and other US Government surplus warehouses. At times, the Weapon System Support Pod (WSSP), a small overnight luggage carrier that attached to the underbelly of the NASA T-38 astronaut training aircraft on overnight cross country flights, became a practical means of transporting surplus equipment and materials for the lab. Dr. Chang Díaz's astronaut training flights would take him near other government facilities and national laboratories where surplus equipment was readily found.

It was clear, though, that in order to strengthen the project and provide it with the credibility it deserved, the VASIMR® team needed to grow, establish scientific clout and achieve institutional diversity. This was especially true at JSC, an institution focused on Space Shuttle operations and training, with only limited research activities in space medicine. Collaboration on the rocket experiment needed to develop at the local level first, as the team's travel budget was virtually non-existent. In the local area, the VASIMR® researchers knew that several top-notch Texas universities had strong programs in plasma physics, including the University of Texas at Austin, Rice University and Texas A&M. The program at UT-Austin was the strongest and included the prestigious Institute for Fusion Studies, a well-known group in plasma theory, as well as the Fusion Research Center, home of the TEXT Experiment, a DoE-funded Tokamak, which had initiated operations in 1980.

In November of 1996, Dr. Chang Díaz and Dr. Squire attended the 38th Annual Meeting of the Division of Plasma Physics of the American Physical Society in Denver, Colorado, where they presented "*A New Facility to Investigate Space Propulsion using Magnetically Vectored and RF Heated Plasmas*," their first formal presentation to the plasma physics community on the new experiment at NASA's ASPL. In that meeting, the "helicon," an RF discharge developed for the plasma processing industry by Dr. Roderick Boswell of the Australian National University and Dr. Francis Chen of the University of California at Los Angeles, was featured prominently. From the technical presentations, Dr. Chang Díaz

4.8 STS-75, February 1996: On the flight deck of the Space Shuttle *Columbia*, Dr. Chang Díaz controls the deployment of the tethered satellite. At 19 km of tether length, the re-flight achieved almost total deployment. However, a high-voltage breakdown of the tether insulation at the base of the deployment mechanism on the Shuttle cargo bay severed the tether. Though the satellite was lost, the experiment fully demonstrated the electrodynamic tether principle, generating nearly 5 kW of electrical power from the motion of the Shuttle and conducting tether through the Earth's magnetic field.

became extremely interested in the helicon source as a possible ionizer stage for the VASIMR®. The high-density plasma generated by these RF sources was exactly what was needed to provide the initial plasma. Earlier designs for the VASIMR® ionizer included the MPD and more conventional microwave discharges. However, the presence of electrodes in the MPD implied a potential material erosion issue which could prove too challenging for a high power plasma rocket. The microwave approach, on the other hand, did not have electrodes, but the equipment was bulky and heavy and would be difficult to package in a flight system. The helicon eliminated both of these problems, as its much lower frequency allowed for compact and lightweight RF power equipment, given the expected advances in solid-state technology.

The Denver meeting was a watershed for the VASIMR® researchers. In addition to identifying and learning about helicons, the meeting marked the beginning of a strong and productive relationship with two important plasma physics groups: The Fusion Research Center of the University of Texas at Austin and the RF group of the Fusion Energy Division at the Oak Ridge National Laboratory (ORNL). Shortly after returning from Denver, Dr. Squire initiated contact with Dr. Wally Baity and Dr. Richard Goulding of ORNL, both of whom he had met in Denver. At about the same time, Dr. Alan Wootton, Director of the Fusion Research Center at UT-Austin, accepted Dr. Chang Díaz's invitation to visit the ASPL.

Dr. Wootton's visit was short, but he was clearly impressed and pleasantly surprised that a project of that scale and scope could be taking place at JSC, an institution hitherto unknown in the world of plasma research. He promised to go back to UT and discuss with his team the possibility of initiating a collaborative relationship. His visit triggered a second visit by one of his close associates, Dr. Roger Bengtson, an experimentalist from the Fusion Research Center, who spent several hours touring the laboratory and discussing the physics of the plasma engine with the Houston team. There were a lot of questions that needed to be answered, but Dr. Bengtson was intrigued by the project and proposed to make arrangements for Dr. Chang Díaz to give a full seminar to the physics faculty at UT. The seminar was to be an informal review and vetting exercise for the VASIMR® system, particularly if Dr. Herbert Berk, a renowned authority on magnetic mirrors and one of the top scientists at UT, would accept the physics premises upon which the engine was based.

Dr. Chang Díaz's VASIMR® engine seminar at UT took place in February of 1998. The seminar was successful in bringing scientific credibility to the concept. After the formal presentation, Dr. Chang Díaz spent most of the day answering questions and describing the research plan he and Dr. Squire had developed. Two UT physicists would take the lead in establishing the ASPL-UT collaboration: Dr. Roger Bengtson, who was now well acquainted with the project, and Dr. Boris Breizman, a research scientist and plasma theorist from Novosibirsk, Russia, who had immigrated to the US several years before. A couple of months after Dr. Chang Díaz's UT visit, Dr. Berk visited the ASPL for further discussions, accompanied by Dr. Breizman. All these interactions had produced generally positive feedback and although the technology was clearly far from mature, no scientific "show stoppers" had been identified.

It was a great start, but the local Houston team still needed to grow. In February of 1997, Timothy W. Glover, a PhD graduate student in the Department of Physics and Astronomy at Rice University, approached Dr. Chang Díaz seeking a position as a research assistant. As he had done at MIT, Dr. Chang Díaz was looking to engage young graduate students from the local Houston universities. Rice was ideal and Tim was an excellent candidate. He was methodical, exacting and well organized and was clearly enthusiastic about the project. Originally from Saskatchewan, Canada, Tim was extremely well trained in the field. He had obtained BS and Master's degrees in Physics at New Mexico State University and the University of Pittsburgh respectively, before transitioning to an additional Master's in Aerospace Engineering at the University of Texas at Austin. He was a hardworking, quiet and easy-going young man, who adapted quickly to support the needs of the laboratory.

Tim's first attempt at a PhD thesis project involved the MPD VASIMR® plasma injector, which Dr. Chang Díaz had acquired from Dr. Edgar Choueiri, the new Director of the Princeton University Electric Propulsion Laboratory. However, the MPD work was short-lived, as the "helicon" RF plasma source was quickly gaining traction as the injector of choice for the engine, resulting in a more natural ionizer stage while also completely eliminating the need for physical electrodes in the plasma. Tim's PhD thesis, supervised by Professor Anthony Chan of Rice University, and Dr Chang Díaz at the ASPL, eventually shifted to obtaining the first measurement of the ion acceleration by ion cyclotron resonance in the VASIMR® magnetic nozzle, a goal which he ultimately achieved. Also in the summer of that year, two UT students, Verlin Jacobson and Robert Bussell, were assigned to the project; their graduate student support materialized from small bits of funding that Dr. Chang Díaz was able to obtain along the way.

4.9. The fully assembled ASPL experiment at NASA JSC and some of the team members at work are (from L to R) Dr. Squire, Dr. Ilin, "Buddy" Goebel and Tim Glover (circa 1997).

The ASPL project had now become better known in the Texas plasma physics community and elsewhere and had caught the attention of young investigators. The team was growing and it was refreshing to see so many young people flocking to the laboratory every day in search of a position in the group. From its early days at JSC, the ASPL became one of the most popular assignments for young NASA interns, who came to JSC with paid summer jobs which would prepare them for potential future positions at NASA. Ironically, however, the management of the NASA JSC Intern Program discouraged JSC interns from participating in the ASPL research in their last tours, arguing that the laboratory was not part of the official JSC organizational structure and NASA positions in the ASPL were not contemplated to open up. It was a sad truth that, even within the tenuous shelter of the Astronaut Office, the ASPL remained an institutional orphan and would continue to be so for all but the last couple of years of its existence at NASA. These realities notwithstanding, several JSC interns forfeited the chance of a NASA job and joined the VASIMR® team anyway, becoming valuable assets to the research. Their hard work helped move the project forward and contributed to the scientific breakthroughs that soon followed.

The popularity of the ASPL also spread well beyond the technical circles and caught the attention of the movie industry. In the fall of 1996, a Walt Disney group approached the team to discuss the potential use of the ASPL as a laboratory setting for the filming of Stuart Gillard's science fiction comedy "*Rocketman*," the hilarious story of computer whiz, but accident-prone, Fred Randall, who is chosen as part of the crew for the first mission to Mars. Fred undergoes rigorous training at NASA, some of which takes place at the Advanced Space Propulsion Laboratory. The cast included actor Harland Williams who plays Randall, William Sadler in the role of Commander "Wild Bill" Oberbeck, Jessica Lundy as Mission Specialist Julie Ford and actors Beau Bridges and Jeffrey DeMunn in supporting roles. One of the training scenes of the unlikely astronaut, shot in the ASPL, shows brief cameo appearances by Dr. Squire and Dr. Chang Díaz.

For the ASPL scientists, it was an extraordinary experience to participate in the filming of a full-length Hollywood movie. During the shooting of the brief ASPL scenes, the production team of "*Rocketman*" completely took over the laboratory for a full week. Close to 100 people associated with the movie descended on the facility, including producers, directors, actors, their entourages and dozens of makeup artists, technicians, wardrobe experts and support personnel. The laboratory was fitted with special lights, props and cameras and at times, the technicians would purposely spill liquid nitrogen on the floor in order to provide an appropriate visual effect. The scene was repeated over and over until the director was satisfied with the take.

To the ASPL team, accustomed to a meager subsistence budget, the financial scale of the filming operation was astounding. There was clearly a lot of money tied up in the production of the movie and, compared to their level of funding, the film project's abundance of resources was strikingly evident. For a full week, the large parking lot outside the ASPL building was full of Disney Productions vehicles and most of the area had been used to set up a gigantic heavy canvas tent as a break room and gathering place for the production team. The tent was fitted with large benches and tables for informal dining and provided an endless supply of snacks, fruits and other edibles. Large diesel generators provided air conditioning. Breakfast, lunch and dinner was catered from far away, so hot meals would be available over the course of the week. Over several days of shooting operations, the special effects teams consumed hundreds of liters of liquid nitrogen, which was delivered to the laboratory by the ASPL liquid nitrogen contractor and paid for by Disney. Liquid nitrogen was an extremely valuable resource that the ASPL team used during normal laboratory operations to chill the cryogenic magnets during plasma shots. To the ASPL researchers, accustomed to exercising great frugality in the use of this resource, it was horrifying to see it poured on the floor of the laboratory just to achieve a special effect.

For their parts in the movie, the producers wanted to pay Dr. Squire and Dr. Chang Díaz. However, while there was no problem with Dr. Squire, as a private contractor, Dr. Chang Díaz, being a civil servant, could not accept pay. The producers also wanted to pay the ASPL for the use of their facilities. Sadly, however, while strapped for operating funds, the ASPL could not accept the payment as the funds would have to go directly to the US National Treasury, never to be seen by the ASPL team. The producers, however, devised a clever way to circumvent the government bureaucracy, by supplying the ASPL with several hundred liters of liquid nitrogen, which was purposely left in the laboratory as unused special effects waste materials to be discarded. The ASPL team duly obliged and "discarded" the fluid through the VASIMR® cryogenic magnets (thus providing cooling to the coils) over the course of several weeks of plasma operations.

By 1998, the core ASPL team had been formed and included a couple of additional investigators: Greg McCaskill, an RF engineer who had been working as a NASA contractor on radar communications, and Dr. Andrew Ilin, a Russian mathematician who worked for Lockheed Martin and transferred to the rocket project in December of 1996. The budget woes remained, however, and the group survived on small bits of funding at the $50,000 level that materialized from time to time, primarily through the Center Director Discretionary Fund (CDDF) Program. Despite the budgetary constraints, however, the scope of the research was far reaching, as some of the external collaborators, particularly the ORNL group, leveraged their work with internal DoE funds. Their team was part of the Fusion Energy Division, led by Dr. Stanley Milora, a visionary MIT-educated plasma physicist, who quickly understood the significance of the VASIMR® technology and the important contribution his team could provide. Dr. Milora agreed that the Mini RFTF experiment at Oak Ridge, although designed to test RF technologies for fusion, was similar in scope and could be used to validate the physics principles at work in the VASIMR®. There were key questions to be addressed, including the efficiency and energy cost of ionizing the feedstock propellant in the first stage, as well as the actual demonstration of ion acceleration by Ion Cyclotron Resonance Heating (ICRH), a theoretical process that had yet to be experimentally observed. ICRH is a common technique for heating plasmas under magnetic confinement fusion, but relies on the particles to be confined long enough within the magnetic bottle. Estimates for the VASIMR® engine, however, indicated that efficient heating could be accomplished without the need for significant confinement, in a sort of "single pass" of the particles as they flew out of the rocket. Such a possibility, if experimentally demonstrated, was very exciting, as it would eliminate the need for multiple particle reflections in the central cell of the device, making the engine considerably simpler.

FROM COMPETITION TO COLLABORATION

Altogether, the collective works of the extended VASIMR® team now spanned several universities, a National Laboratory and one NASA center, JSC. The word, however, was rapidly spreading throughout the rest of NASA with a curious mix of reactions; some enthusiastically supportive, others not. After all, the project had been shut down by NASA years before while at MIT and the rebirth of the research was a bit surprising to some. One of the NASA centers where the news of the ASPL quickly resonated was the Marshall Space Flight Center (MSFC) in Huntsville, Alabama, home of former rocket scientists Wernher von Braun and Ernst Stuhlinger. They were part of the famous team of German scientists brought to the US after World War II to jump start the nation's early rocket program, and who ultimately developed the rockets for the Apollo Moon missions.

During all these years, and probably distracted by much larger programs in chemical propulsion, MSFC had remained on the sidelines of plasma propulsion. While it remained a NASA center of excellence for rocket propulsion, they had been dismissive of electric propulsion, concentrating instead on large chemical boosters and the supporting engineering associated with the Space Shuttle main engines. Nonetheless, the popularity of the VASIMR® project had begun to pique their interest. A small MSFC group had followed the project and endeavored to pursue their own version, a virtual carbon copy of the VASIMR® engine, which of course they called by a different name and, in a clear display of scientific naiveté, claimed it could become a fusion rocket.

To the VASIMR® team the news of the MSFC initiative was both ominous and disheartening. It was evident that Huntsville, the home of the rocket scientists of Apollo fame, had a name to live up to. The news from Houston had awakened a small group of newcomers to the field who, with little background in either plasma physics or electric propulsion, saw an opportunity to make history again. To the VASIMR® team, it felt as if their near twenty-year journey through the physics of plasma rockets was being purposely dismissed or ignored and conveniently renamed in order to be transplanted. A new threat had appeared that preyed on the perceived programmatic illegitimacy of the Houston propulsion research facility and the need to bring the project to what could be viewed as its rightful place.

On June 24, 1997, two visitors from MSFC came to tour the ASPL: Mr. John Cole, Manager of the Revolutionary Propulsion Research Project Office and Mr. William Emrich, an MSFC engineer and graduate student at the University of Alabama, Huntsville. Unbeknown to the ASPL team, they had been discussing a conceptual fusion rocket, called the Gas Dynamic Mirror (GDM), with Terry Kammash, a Professor of Nuclear Engineering at the University of Michigan. Professor Kammash had proposed the concept years before as a fusion reactor. Unfortunately, the GDM was a system that, while theoretically defensible, inherited many of the technological and physics challenges that had resulted in the demise of the magnetic mirror fusion program in the 1980s.

The MSFC engineers, however, did not appear to have become sufficiently familiar with these prior studies and wanted to build an experiment at MSFC to prove the concept. Their visit brought mixed emotions to the ASPL team. On the one hand, it was refreshing to see the apparent enthusiasm of the visitors and it was not a time to squelch new initiatives. On the other hand, in the opinion of the ASPL team, the GDM was a highly speculative and technologically naïve concept. Moreover, their proposed experiment was scientifically irrelevant, for it was quite evident that the parameter space of such an experiment would not approach the operational physics regime of the GDM. Curiously enough, the magnetic topology of their proposed experiment was nearly identical to that of the ASPL VASIMR® device; therefore, the MSFC management argued that the ASPL hardware needed to be moved to Alabama.

Coming from a propulsion center of excellence, this threat could not be ignored and had to be addressed head on. Thus, as the ASPL laboratory started its new life in Houston, yet another battle for acceptance and survival began to brew. A clear undercurrent of the well-known inter center rivalry that has existed for decades between the Johnson and Marshall Space Flight Centers could be felt during the meeting. Both teams found themselves silently witnessing the "worst" of both worlds: two NASA field centers, one with the scientific expertise but lacking the programmatic mandate, the other with the mandate but without the expertise. It was a classic byproduct of NASA's forced technology compartmentalization of its field centers and it was uncomfortable for the ASPL scientists to sense a potential takeover in the making. The JSC team instead proposed a JSC-MSFC collaboration to continue to move the VASIMR® engine forward at the ASPL.

The VASIMR® team knew that a direct quest for a fusion rocket was not the correct forward strategy. It was best instead to seek a more near-term "precursor." They had learned, over more than a decade of experiments, that realistic parameters for a plasma rocket, driven by solar- or nuclear-electric fission, were already compelling enough to support a robust solar system exploration mission. Early plasma rockets, such as the VASIMR® engine, while still technologically difficult, were far easier to achieve than fusion.

They would spur the technology forward and provide expertise, along with an early operational payoff. These early precursors, in due time, would provide a natural bridge to fusion. As the VASIMR® team saw it, the path to fusion implied a long and challenging journey, with the best strategy being numerous small steps rather than a few giant leaps. More than anything, it was important to not move backwards.

Rumors of a potential relocation of the VASIMR® project to MSFC began to circulate shortly after the MSFC visit. Fortunately, the laboratory was now fully established at JSC and the experiment was already producing useful scientific data; both of these facts contributed to reducing the impetus for the move. Moreover, Mr. John Cole became a strong supporter of the VASIMR® project and, relocation notwithstanding, Dr. Chang Díaz was determined to build a strong relationship with MSFC. In 1999, two MSFC engineers, Ms. Carol Dexter and Mr. Greg Chavers, were assigned to work with the ASPL team in Houston and became valuable contributors to the technology. Mr. Chavers became a PhD Student at the University of Alabama, Huntsville, and initiated his thesis work at the ASPL under Dr. Chang Díaz's supervision. Through this relationship, along with his appointments at Rice University and the University of Houston, Dr. Chang Díaz was named Adjunct Professor of Physics at the University of Alabama, Huntsville.

4.10 The first VASIMR® Workshop took place in 1998 at JSC's ASPL in Houston and was attended by investigators from NASA JSC, GSFC, MSFC, the University of Texas at Austin, the Oak Ridge National Laboratory, Rice University and the University of Maryland.

Another important ally at MSFC was Dr. Stephen Rodgers, who became head of the Propulsion Research Center at Marshall. He was an excellent manager, who was able to see past inter-center rivalries and quickly appreciate the value of the VASIMR® system to NASA and the nation. The ASPL team had not been formed by decree, but had grown organically over many years from methodically addressing the physics unknowns. Recognizing these facts, Dr. Rodgers endeavored to insulate the project from the bureaucracy and continued to promote the research, keeping at bay multiple initiatives to relocate the laboratory to MSFC. His efforts paid off. In a memo addressing a "Review Requirement to retain propulsion function at JSC versus moving it to MSFC or SFOC (SRR Action #136)," dated January 8, 2002, the relocation issue was resolved and closed.

The ASPL had achieved a technically impressive program and to guide it, the team created the VASIMR® Workshop, an annual gathering in Houston at which all the investigators would present their results and the group would outline objectives and goals for the following year. The first of these workshops took place in March of 1998. This meeting succeeded in establishing a physics baseline for the VASIMR® engine and, more importantly, the key knowledge gaps and questions still to be addressed.

REFERENCE

1. *Rapid Mars Transits with Exhaust Modulated Plasma Propulsion* NASA Technical Paper 3539, May, 1995.

5

The Breakthroughs

The incorporation of the helicon ionizer, together with a number of other refinements, resulted in a new US patent (6,334,302 B1) for the rocket, which was filed in June of 1999 and granted in January of 2002. In contrast to most laboratory helicons at the time, where the plasma remained in place in the RF cavity, the VASIMR® helicon ionizer was magnetically asymmetric and the plasma was required to flow continuously from one end to the other. With this modification, the helicon ionizer became the first stage of a three-stage VASIMR® engine, with the RF booster and the magnetic nozzle as second and third stages respectively. The transition from one stage to the next had to occur seamlessly and efficiently, with no instabilities or other adverse effects. The physics tying these three elements together were not well understood in the late 1990s. Although many theoretical models were available, their experimental verification was a difficult work in progress, with many unresolved issues. The ASPL team sought to address these unknowns with a series of experiments.

To accomplish this onerous set of tasks, the collective work of the research team in the late 1990s followed a comprehensive plan that had been laid out by the group during the first VASIMR® workshop back in 1989. This plan was reviewed and updated annually. The most critical physics unknowns were:

1. Could the engine achieve efficient and complete ionization of the propellant?
2. Would the "single-pass" acceleration of the ions in the rocket's second stage be possible and, if so, would it be efficient?
3. Once accelerated by the magnetic nozzle, could the plasma detach from the magnetic structure tied to the rocket to provide useful thrust?

It was also important, given the budget and infrastructure limitations, coupled with the inherently high power levels of the VASIMR® engine, to implement a thrust measurement without the traditional thrust stand used in low power electric thrusters. The thrust measurement had to be simple, inexpensive and, above all, accurate and reliable. It also had to be compatible with the configuration of the Advanced Space Propulsion Laboratory (ASPL) experiment, where the vacuum system was integrated into the body of the rocket itself. The implementation of a thrust measurement without the use of the standard thrust stand employed in low power electric rockets was a necessary but controversial departure

© Springer International Publishing Switzerland 2017
F. Chang Díaz, E. Seedhouse, *To Mars and Beyond, Fast!*, Springer Praxis Books,
DOI 10.1007/978-3-319-22918-8_5

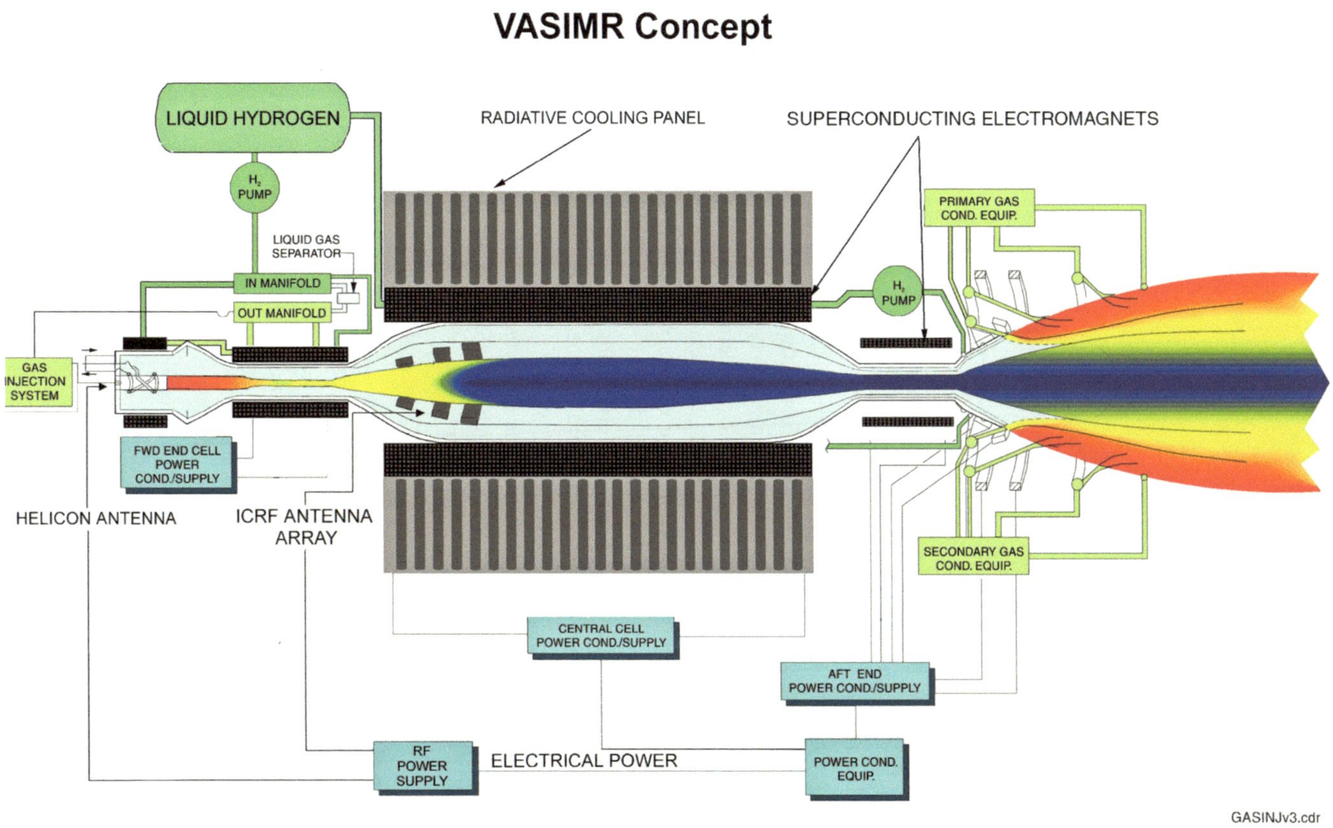

VASIMR Concept

SUPERCONDUCTING ELECTROMAGNETS

RADIATIVE COOLING PANEL

PRIMARY GAS COND. EQUIP.

SECONDARY GAS COND. EQUIP.

H₂ PUMP

AFT END POWER COND./SUPPLY

POWER COND. EQUIP.

CENTRAL CELL POWER COND./SUPPLY

ELECTRICAL POWER

LIQUID HYDROGEN

H₂ PUMP

LIQUID GAS SEPARATOR

IN MANIFOLD

OUT MANIFOLD

ICRF ANTENNA ARRAY

FWD END CELL POWER COND./SUPPLY

RF POWER SUPPLY

GAS INJECTION SYSTEM

HELICON ANTENNA

GASINJv3.cdr

5.1 The new VASIMR® engine concept incorporated the helicon ionizer as its first stage. The improved design eliminated the need for physical electrodes in contact with the plasma.

from tradition and elicited a great deal of philosophical opposition. It was important for the VASIMR® team to address this issue on solid physics grounds. This demonstration was achieved later in the program with a peer reviewed "blind test" of the ASPL thrust sensor carried out at the University of Michigan, where the sensor's measurements on a Hall thruster were compared with those obtained from the same engine in a traditional thrust stand. The variation was shown to be less than one percent. We address this topic in greater detail in Chapter 7.

THE HELICON PLASMA SOURCE

In 1996, the group delved deeply into the field of helicon plasma sources. Helicons had begun to be used in the microelectronics industry to efficiently produce homogeneous plasmas for etching integrated circuits, with much greater precision than that obtained by traditional chemical etching methods. The process was also considered to be much cleaner and more environmentally friendly due to the lack of etching chemicals that ultimately would have to be washed away and discarded, with a high environmental impact. However, to the VASIMR® team, as their recent patent indicated, the helicon had not been broadly considered as a propulsive device. Nonetheless, quite a few papers soon began to surface in the US and abroad that capitalized on the popularity of the ASPL project and presented numerous variants of the helicon as a propulsive device, both with and without a VASIMR®-like multi-stage configuration. The ASPL team saw these as positive developments, indicating a general interest within the research community to seriously explore the fundamental physics on which the VASIMR® engine is based.

A typical helicon plasma source consists of a cylindrical assembly, where an electrically insulating tube – usually a high temperature ceramic – is fitted with an external radiofrequency (RF) antenna. This assembly is placed inside the evacuated bore of a solenoidal magnet. With the magnet energized, feedstock gas is injected at one end and is turned into dense plasma by the action of the RF waves. The plasma, in the case of the VASIMR® engine, senses a pressure gradient in the presence of the axial magnetic field and flows along the field lines to the exhaust end of the tube, producing a plasma jet. The plasma temperature within the discharge could be up to tens of thousands of degrees, well above the melting point of even the toughest material. Fortunately, the magnetic field acts as an invisible insulating liner that guides the plasma along and partially shields the ceramic wall of the tube. While the tube does get hot, mainly from the ultraviolet radiation emanating from the plasma, it can maintain an acceptable temperature with proper external cooling.

The physics of helicons is based on the stimulation of natural waves in a magnetized plasma by means of a suitably designed RF antenna. The RF field from the antenna drives natural electromagnetic waves in the plasma. These so-called "helicon" waves in turn induce ionization of the target gas by delivering kinetic energy to free electrons existing naturally in the gas. Under the proper conditions, the waves resonate with the electron gyromotion: the rotating "corkscrew" motion of a negative charge as it orbits about an externally imposed magnetic field. Thus, helicon waves are often described as circularly polarized. While sufficiently well understood to allow technological applications, the physics of these devices continues to be the subject of extensive study.

As the electrons gain energy from the waves, they collide with neighboring atoms in the gas, freeing additional electrons which also gain energy and, in turn, create additional collisions and hence more electrons. This process results in an ionization cascade that turns the gas into a plasma. Helicons are very effective at obtaining high plasma densities – as compared with other ionization schemes – with fairly simple configurations. Many different ceramics are used for the dielectric tube and many types of antennas are also possible, depending on the application, the power level and the required plasma density. In addition, the magnetic field topology, its strength, as well as the chosen frequency, are important factors in determining the efficiency of the device; namely how much energy is required to produce an electron-ion pair. This latter quantity, called the "ionization cost," expresses a very important design parameter for efficient rocket operation.

The ionization cost is the energy "tax" that must be paid to manufacture a plasma from neutral gas. The actual value of this quantity depends on many factors, including the type of device employed and, of course, the feedstock gas. Some propellants, such as argon and xenon, are relatively "energy cheap" to ionize, while others, such as hydrogen and helium, are not, as their electrons are more tightly bound to the atomic nucleus. The ionization cost is measured in units of energy – electron volts (eV) to be specific – and for efficient rocket operation, it must be kept as low as possible compared to the total kinetic energy given to the ion in the exhaust. Generally, high power and high specific impulse (I_{sp}) devices, operating at high exhaust energy, can tolerate a higher ionization cost than low power devices, operating at relatively low ion energy.

An important feature of helicons as efficient plasma sources for high power electric rockets is their high density. With the correct combination of power and magnetic field, these devices exhibit a very favorable high density transition, which also tends to drive the waves conveniently deep into the discharge. The frequency of choice generally lies between the electron and ion cyclotron frequencies. In the early experiments, the ASPL team was able to trigger the high density transition in the experiment at power thresholds of several kilowatts. In these high density modes, the plasma would jump in density over nearly an order of magnitude.

In 1997, the ASPL team began to study helicon discharges in a variety of gases, including argon, helium, hydrogen, deuterium, nitrogen and xenon, quickly recognizing the wide range of potential propellants that could therefore become available to the VASIMR® engine. The flexibility of using multiple propellants in a single engine implied that an additional method of constant power throttling (CPT) could be available; namely, changing the atomic mass of the propellant to increase or decrease the I_{sp} at the same power, with corresponding changes in thrust. In this way, one could also envision different versions of the same engine, being optimized for cislunar space with lower I_{sp} and higher thrust requirements, and for deep space where the optimal I_{sp} could be much higher. These features provide a great deal of operational flexibility to the mission planners.

The first VASIMR® helicon ionizer, a simple double twist antenna saddled on a quartz tube, was designed in early 1997 with the help of the Oak Ridge National Laboratory (ORNL) team, who were also experimenting with helicon plasmas in their Mini RFTF (Radio Frequency Test Facility) experiment. The team included Dr. Wally Baity, Dr. Mark Carter, Dr. Richard Goulding, RF engineer Mr. Glenn Barber and RF Master Technician Mr. Dennis Sparks. The device demonstrated the feasibility of using such a helicon plasma source for the VASIMR® engine. The first ASPL helicon ionizer system was cooled by de-ionized water and was able to run at 1 kW for over 15 minutes. The system operated with helium gas at flow rates of approximately 0.5 mg/s and frequencies of 8 MHz and 13.56 MHz. These choices were driven purely by the available laboratory surplus equipment, which the ASPL team temporarily borrowed from Oak Ridge. For good data collection, because of the pumping limitations in the modest experimental hardware, measurements had to be obtained in the first few milliseconds of the discharge, while the pressure remained at or below 5×10^{-5} torr, in order to avoid excessive neutral gas build-up in the system.

The ASPL helicon experiments were highly successful in achieving a stable high density discharge that could form the basis for further work on the rocket. In the year 2000, the ASPL team received the Rotary NASA Stellar Group Award for their achievements in high density helicon design and operation. The recognition was received by Dr. Jared P. Squire in representation of the ASPL team and the award brought more than internal satisfaction to the VASIMR® team. It made the group better known in the NASA Houston space community which, while accustomed to human space flight and operations, was not known for its research in rocket propulsion. This was true rocket science, taking place at none other than the home of the astronauts, and this new technology had a potential to disrupt traditional space transportation. There was a certain sense of the "cowboy" mystique on these ASPL iconoclasts – daring to work against the rigid institutional grain – that fitted very well with the early implausibility of the Kennedy Moon shot.

There is a comic historical note that ended the life of the first helicon that year and prompted the ASPL team to improve on the design. The power feedthrough for the unit was undersized and had been failing, but ASPL electrical engineer Mr. Greg McCaskill had continued to nurse it along with multiple fixes, as a suitable replacement was beyond the budget capability of the team. However, the feedthrough finally failed just as the team attempted to demonstrate the operation of the engine to mesmerized Hollywood film Producer/Director James Cameron, who had arrived to gather material for a new space movie he was developing and was eagerly anticipating a view of the awesome plasma discharge. It was to no avail and Dr. Chang Díaz vividly recalls the thick droplets of sweat that covered the pale-white forehead of engineer McCaskill as the famous movie producer, and his considerable entourage, looked over his shoulder with great anticipation. Since that moment, the VASIMR® team has jokingly guarded against the "Cameron Effect," defined as any critical failure path in the engine design that could cripple the operation at the most undesirable moment.

The massive Continental Electronics FRT-86 RF generators, brought to Houston from MIT, were extremely old and unwieldy. The ASPL team, together with experts from ORNL, worked for many months to bring one of these systems back to life, but with only limited success. During this initial period, Mr. Greg McCaskill had identified a 50 kW pulsed radar transmitter that might be available from the Orbital Debris Program at NASA JSC. With some work, the device could be tuned to 13.56 MHz and be modified to run continuously, and hence become suitable for helicon operation. The team requested, and was granted, a temporary loan of the hardware, which Mr. McCaskill installed on the experiment. This hardware enabled the first steady state helicon discharges in the device. The team's advances were formally presented in the summer of 1998, in two invited scientific papers at the Open Systems Conference in Novosibirsk, Russia, which were published in 1999 in the journal *Transactions of Fusion Technology, 35, 87-93 (1999)*.

While the double twist ionizer produced healthy plasma discharges, at 10^{17} particles per m^3, the density was still below the range of interest for the VASIMR® engine application.

5.2 The first helicon antenna was designed in 1997 by the combined ASPL and ORNL teams. The antenna, a simple, water-cooled, double twist design, was integrated into the experiment that same year. Experiments with this configuration provided the foundation for improved designs that came later.

At the same time, the ionization cost was still prohibitively high. Nonetheless, the ASPL team continued to explore the parameter space and introduced step-by-step modifications to the system design that, over time, produced major improvements in plasma density and ionization cost. The present day VASIMR® engine ionizer stage produces plasma at an ionization cost below 100 eV and is capable of plasma densities in excess of 10^{20} particles per m^3, a 1000-fold increase over the earlier experiments.

Despite the hardware constraints, the results of these early experiments were promising enough to continue to focus the team's attention on further improvements of the ionizer design, with a more optimized partial-turn, water-cooled, helical-type antenna. The new antenna and power feedthrough were more robust and were able to handle higher power levels. Also at this time, the VASIMR® team began to consider the use of compact solid-state RF amplifiers. These had also been proposed by the ORNL collaborators and were based on high power MOSFET technology. A one kW demonstration unit was actually manufactured at ORNL and brought to the ASPL for testing. At the time, these modifications to the ASPL device were already capitalizing on technological advances that were coming of age in the late 1990s, which had begun to demonstrate the light weight (~1kg/kW) power conversion capability more relevant to space flight applications of the VASIMR® engine. It would not be until 2005, however, as the ASPL transitioned into the Ad Astra Rocket Company, that serious work would actually be undertaken with solid state RF equipment, through a new corporate collaboration with Canada's Nautel Ltd.

More progress came quickly, as the Mini RFTF team at ORNL already had access to a fairly significant supply of idle Department of Energy (DoE) RF equipment from the

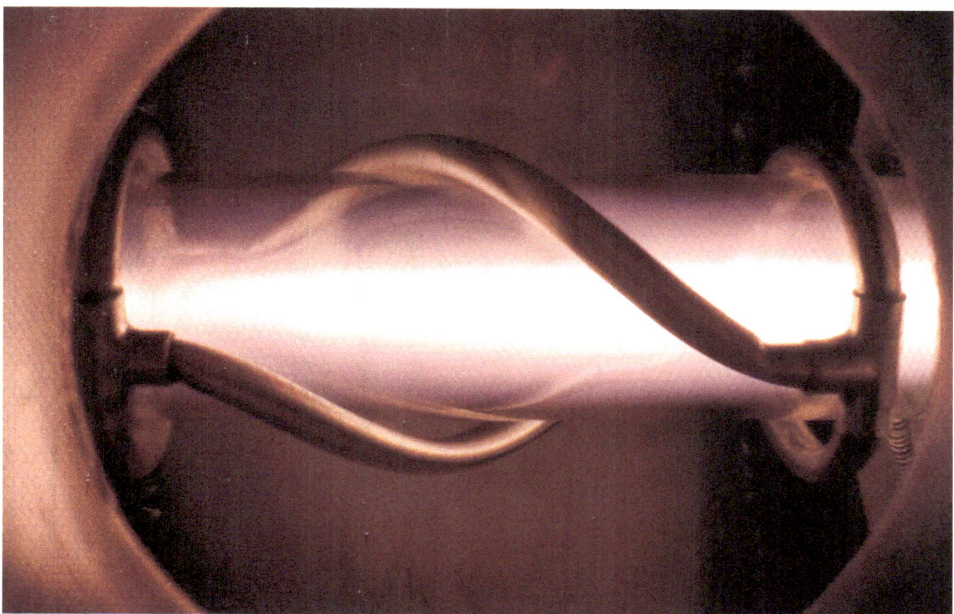

5.3 The helicon ionizer utilized in the late 1990s produced dense stable plasma with a variety of gases and made the plasma rocket electrodeless.

fusion program and quickly identified additional hardware, which enabled increased power and flexibility. An FRT-85 RF transmitter had been included in the hardware shipped from MIT, although this unit lacked a high voltage power supply. Fortunately, this was another component available from ORNL and was also loaned to the ASPL in Houston. The ORNL engineers, led by Mr. Glenn Barber and Mr. Dennis Sparks, installed and tuned the new unit and incorporated the system onto the ASPL experiment. The new configuration was called the VX-10, a nomenclature conceived by Dr. Squire to name the various evolutions of the VASIMR® technology. In this way, the "V" stood for VASIMR®, the "X" for "experimental" and the numeral indicated the power level. Accordingly, VX-10 stood for "VASIMR® experiment at 10 kW". The 10kW power limitation was imposed by the helicon hardware and not by the transmitter, which was capable of much higher power. Following this nomenclature, the designation "VF" would stand for VASIMR® flight, a nice prospect to plan for. The VX/VF nomenclature has been adopted with all the subsequent experimental prototypes at the ASPL and later at the Ad Astra Rocket Company. These include the VX-50 in 2004, VX-100 in 2006 and VX-200 starting in 2008. In 2009, Ad Astra initiated the conceptual design of a proto-flight 200 kW engine, designated as the VF-200-1 and in 2015, following the award of a major NASA contract, the company began the development of the VX-200SS (for steady state). This engine features a new thermal management system that will extract the heat radiated by the plasma to the rocket's casing. With this system, expected to be demonstrated in 2018, the VASIMR® will be able to operate at high power levels indefinitely. This, so-called "thermal steady state operation," is considered to be the last remaining engineering hurdle to be overcome before the VASIMR® engine is ready for its first flight into space.

THE TEAM LOOKS SKYWARD

In mid-1997, Dr. Chang Díaz was assigned to STS-91, his sixth mission to space. This time, the Space Shuttle *Discovery* was to link up with the Russian space station MIR and deliver supplies and new equipment to the aging outpost. The most important scientific objective of the mission, however, was the maiden flight of the Alpha Magnetic Spectrometer (AMS). This was an ambitious particle physics experiment designed to search for evidence of dark matter, as well as naturally occurring particles of antimatter and other high energy elements – such as cosmic rays – which could be detected and measured before they were affected or otherwise transformed by the Earth's atmosphere. The payload was a large magnetic mass spectrometer, an array of sensitive detectors arranged at the base of a powerful and carefully aligned permanent magnet. The magnetic cavity was large, about the size of a Jacuzzi, and, along with the detectors, the total payload included an array of processors, computers and an assortment of electronics weighing several tons. The magnetic field was carefully designed to provide a very uniform force to charged particles, deflecting them depending on their charge/mass ratio. This enabled the particles to be counted and cataloged.

The AMS program was led by MIT physics Professor Samuel C. C. Ting. In 1976, Professor Ting shared the Nobel Prize in Physics with Burton Richter of the Stanford Linear Accelerator Center for their discovery of the fourth "charmed" quark, one of the six fundamental particles in the physics Standard Model of the structure of matter. During his graduate student years at MIT, Dr. Chang Díaz had become aware of the news but had never had a chance to meet Professor Ting, even though the Professor's office at the MIT Laboratory for Nuclear Science on Vassar Street in Cambridge was just across the street from his graduate student office in building 38. Professor Ting led a team of more than 500 scientists at the European Center for Nuclear Studies (CERN) in Geneva, Switzerland, and had a reputation for his disciplined approach to science, particularly for his leadership of a large team of scientists working together internationally for a common goal. He was an extraordinarily persistent and uncompromising leader and ran his projects with an iron fist.

One of the first meetings of the STS-91 crew with Professor Ting occurred in 1997, as the astronauts were becoming acquainted with the payloads they would be responsible for. As the mission's designated Payload Commander, Dr. Chang Díaz became intimately familiar with all the scientific aspects of the mission. However, as a physicist, he had become enthralled by the significance and complexity of the AMS experiment and was determined to do all he could to make it successful. His concerns became evident early in the training when he noted that the payload had no provision for crew interaction from on board the Shuttle. The presumption of the NASA payload integrators was that the ground control team would take care of all of the commands and the involvement of the crew, other than to turn the unit on or off, was not necessary.

Over the years at NASA, Dr. Chang Díaz had always marveled at the reticence of the NASA payload community to significantly involve the crew in the conduct of experiments. A somewhat natural aversion had grown in human space flight between the operations and science communities. The early astronauts, who were trained pilots, tended to view science as secondary, preferring instead to focus on flying the vehicle. Such priorities were totally understandable, coming from those who were prepared to lay their lives on the line. After all, the early spacecraft were still experiments in and of themselves. But the relative indifference to science, a remnant of Mercury and Apollo, still persisted in the Shuttle program. It was neatly summed up by a tongue-in-cheek comment from one of the early Space Shuttle flyers in the 1980s, who said: "The only reason to fly the Space Shuttle is to land it…"

Things began to change in human space flight, however, as more technology, reliability and expertise was gained. The nature and ethos of the astronaut was also changing. Over the years, Dr. Chang Díaz had promoted the evolution of the astronaut corps from the traditional flyer mindset to that of the "operational scientist;" a front-line investigator, capable not only of flying the spacecraft, but also of judiciously using the machine and its on-board systems as a scientific tool. He had seen plenty of payload and system failures, both in his own space missions and those he supported from the ground, that could have been resolved by enabling even a measured amount of crew interaction capability. In one of the payload integration meetings at JSC, Dr. Chang Díaz made a plea to Professor Ting to consider the possibility of enabling the crew to have some limited level of on-board commanding capability for the AMS, just in case.

Happily, Professor Ting was receptive to the idea and authorized the development of a computer program for the astronauts to interact with the AMS payload through their portable computers aboard *Discovery*. It was a fortunate decision, as early in the mission *Discovery* developed a critical failure at its primary KU band communications link that, although totally unrelated to the AMS payload, rendered it unable to transmit data or receive uplinked commands through NASA's geostationary satellite network, the principal mode of communication with the AMS. The glitch threatened to severely limit the capabilities of the science that could be conducted during the mission. After multiple failed attempts by the crew to repair the KU link, the back-up program for AMS onboard communication was activated by the crew through its onboard computers. For the rest of the mission, the crew link became the life line of the experiment, enabling the reconfiguration and in-flight optimization of the AMS instruments and, more importantly, collecting the copious amounts of experiment data on mass storage devices onboard, for later analysis on the ground. The experiment was thus saved and the collected data allowed the AMS team to make a convincing argument for the second mission of the instrument. This time, it would be a permanent external payload on the main truss of the International Space Station (ISS).

The success of AMS-1 paved the way for AMS-2, currently on the ISS, and began a strong friendship and scientific relationship between Professor Ting and Dr. Chang Díaz. After the mission, Professor Ting and several of his key investigators from CERN visited the ASPL and maintained their scientific ties with the plasma rocket project, providing a solid physics sounding board for the experiments being conducted at the ASPL. Years later, after the formation of Ad Astra Rocket Company, the engineering team from Scientific Magnetics, the company who developed the AMS-2 superconducting magnet, also developed Ad Astra's highly successful superconducting magnet for the VX-200 rocket prototype. Its unique cryogen-free design has operated flawlessly since 2009 and has enabled demonstrations of the most advanced VASIMR® engine at power levels in excess of 200 kW. In 2017, the unit is still in operation, having been modified to support the long-duration firings required in the next set of high power VASIMR® experiments on the VX-200SS device.

5.4 STS-91, June 1998: On the Russian Space Station MIR, Dr. Chang Díaz tries some old Latin American tunes on Russian Commander Talgat Musabayev's guitar. "It took some getting used to," he recalls. "In weightlessness, the brain has to re-calibrate the inertia of the strumming hand and the movement on the frets." On STS-91, the Space Shuttle *Discovery* docked with MIR for the last joint flight before the initiation of the ISS Program.

By 1998, the ASPL team had already begun conceptual designs for a flight engine, the VF-10, operating at a modest 10 kW power, an imposed constraint driven by recognition of the known limitations of solar arrays of the day. The system was proposed to provide primary propulsion to a small solar-electric space tug concept, also known as the Radiation Technology Demonstrator (RTD). The RTD was a free flyer demonstrator proposed in collaboration with NASA JSC that featured both a xenon Hall thruster and a VASIMR® engine for primary propulsion. It was designed to explore the technology of a radiation-hardened spacecraft, capable of deploying a string of secondary satellites as it spiraled through the Van Allen Radiation Belts. The satellites would be equipped with radiation measurement sensors, which could probe the radiation environment within the Belts and conduct in-situ measurements of the radiation environment over a period of one year. It would then return to low Earth orbit (LEO) for retrieval by the Shuttle orbiter.

Such a system could be of great interest to NASA and commercial satellite operators. In addition, the solar-electric spacecraft could provide the basic building block design for higher power, low cost transport space tugs, operating between LEO and points in cislunar space, such as the Lagrange points. These strategic locations were being considered by NASA and other space-faring nations as suitable for establishing supply depots and staging areas for future missions to Mars and beyond. These considerations prompted important in-depth system studies, such as the point design for a 10 kW engine and a concept for a 50kg super-critical flight hydrogen tank designed in collaboration with Lockheed Martin's Michoud Facility in New Orleans. The formulation phase of the study was conducted in collaboration with the NASA Goddard Space Flight and Glenn Research Centers.

Nevertheless, while the technology objectives of the RTD were clearly defined, the science objectives of the radiation mission could not be fully agreed upon. As a result, the project languished and became lost among the myriad concepts being discussed as NASA struggled to chart a new path for deep space exploration. The experience gained in RTD, however, provided the VASIMR® team with its first look at an integrated flight system. This would prove to be extremely useful in defining another proposed experiment later on: the test of the VASIMR® engine on the ISS.

The RTD experience helped the ASPL team focus on the key experiments that needed to be conducted to lead quickly to a flight demonstration. Starting in 1999, and within the capabilities of the laboratory equipment, the team focused on studying the RTD characteristics in the VX-10 laboratory experimental set up. The existing magnets were operated in such a way as to match the proposed flight demonstration field profile and the research team began the implementation of additional diagnostics to reproduce, in the VX-10, the thermal characteristics of the RTD. Additionally, the group began to experiment with a high mirror field downstream of the helicon antenna, with tantalizing results. Also in that year, the ASPL laboratory became certified to operate with gaseous hydrogen and deuterium, in addition to the traditional laboratory gases that had been used, such as argon and helium. High density plasma discharges with hydrogen were obtained in these experiments.

Along with the experimental advances, the ASPL team also made rapid progress in the theoretical and simulation fronts. A particle code, developed in-house, was integrated with EMIR, a plasma code existing at ORNL, enabling the numerical study of the ion cyclotron resonance heating (ICRH) process with more realism. At the same time, the team's collaboration with the University of Maryland produced a fast, robust, user friendly, optimization code to simulate variable I_{sp} trajectories. With these tools in hand, a parametric study of fast nuclear-electric propulsion (NEP) missions to Mars was completed, which verified and expanded upon the original work done in 1994.

Perhaps in some indirect way, the trajectory studies with VASIMR®-like exhaust modulation had prompted NASA to promote the development of more advanced computational models, which could incorporate the optimization of a propulsion system that operated continuously under schemes such as CPT. These trajectories had to account for the complex gravitational effects introduced by a multi-body problem, such as in a mission to Mars, simultaneously combining the gravitational fields of the Earth, Moon, Sun and Mars. The usual approach to solving these "multi-body" trajectories involved approximations, the first of which simplified the problem by partitioning the flight path into three-dimensional regions of space, called "Spheres of Influence (SOI)." In this way, when the ship was near Earth, it would be predominantly within the Earth SOI and all other gravity effects would be ignored. Likewise, in heliocentric space the Sun SOI would rule and finally, during approach to Mars, the Mars SOI would prevail, with all other effects being ignored.

With these approximations, the resulting trajectories were a patchwork of highly iterative and computer intensive solutions, a trial and error approach where convergence was often difficult to achieve. What was needed was a new code that could optimize the propulsive profile throughout the entire trajectory, transitioning seamlessly from one SOI to the other. Such a program was under development by Dr. César Ocampo, a young Professor of Aerospace Engineering at the University of Texas at Austin. The code, known as Copernicus, was still evolving with NASA support as Dr. Chang Díaz initiated a collaboration with Dr. Ocampo. The goal of this collaboration was for the VASIMR® team to gain rapid insight on a number of missions and to help Dr. Ocampo perfect his program.

The collaboration with Dr. Ocampo turned out to be extremely useful, generating rapid VASIMR® trajectories to Mars, Jupiter and Pluto. The more primitive codes developed by the ASPL served as benchmarks, against which the new code could be measured and the results compared. These exercises led to the rapid generation of new trajectory studies that quickly reaffirmed the importance of exhaust modulation in order to achieve optimal Mars transits with maximum payload and with minimum time and fuel. In addition, the VASIMR® team explored operational abort scenarios that could provide attractive safeguards to human crews in deep space missions in the event of unforeseen failures. Over the years, the Copernicus code evolved into a robust computational tool for electric propulsion and became the "NASA standard" code for calculating continuous, low thrust trajectories. The VASIMR® team is proud of its early contribution towards the development of this important instrument.

TEAM CONSOLIDATION AND INTERNATIONAL EXPANSION

In 2000, the ASPL operation with light gases developed further. High-density hydrogen and deuterium plasmas continued to be produced and a high magnetic field capability (greater than 1 Tesla) was achieved. The experiment user interfaces were also refined, enabling the investigators a faster turnaround and better access to the diagnostics data. At this time, with the experience of the RTD study, the team began a collaboration with Dr. William Schwenterly of the ORNL and Dr. Christopher Rey of the DuPont Corporation, aimed at designing the first flight-like, high temperature superconducting magnet for the VASIMR® engine. This project also promoted discussions with other research groups, particularly Dr. Jentung Ku and the low temperature team of NASA's Goddard Space Flight Center. The group had been working on cryocooler and heat pipe technology, an area of great importance to address the thermal management requirements of the engine. By this time, the size of the VASIMR® team had grown to about 50 scientists, engineers and several graduate and undergraduate students.

By 2001, the ASPL budget requirements had grown to about $1.5 million per year. However, at just close to $1 million, the funding resources were barely able to keep up with the growth of the project. The team had also begun to leverage resources through collaborations from synergistic research by other groups worldwide, as relationships with investigators as far away as Australia, Sweden, Russia and Japan had been established. Anticipating the potential future involvement of young researchers from his native country of Costa Rica, Dr. Chang Díaz also nurtured a relationship with that nation's newly formed Center for High Technology. The Center provided trained manpower in the form of engineering graduates, who came to the ASPL laboratory for periods of six months to one year in NASA unpaid internships, but with travel and subsistence costs furnished by the Costa Rican government. These young researchers added much value to the project in areas such as plasma diagnostics, control systems and superconductivity. Years later, some of these youngsters would become key employees of the Costa Rica subsidiary of a new American company based in Texas by the name of Ad Astra Rocket Company. The subsidiary facility would be located near the city of Liberia in the country's northwestern province of Guanacaste. In 2006, the establishment of the Costa Rica subsidiary of Ad Astra Rocket Company brought a significant amount of Costa Rican private investment to the US parent company and these funds contributed immensely to the maturation of the VASIMR® technology in the United States. All of these developments put to rest the remarkable assertion of a high level NASA official in Washington, who had once asked Dr. Chang Díaz: "Really Franklin, what could Costa Rica possibly have to offer NASA?"

Prejudices notwithstanding, the ASPL team continued to build the relationships and scientific collaborations with international groups in Australia, Japan, Sweden, and others in Europe and the Americas. A collaboration with the Alfvén Laboratory of the Swedish Royal Institute of Technology produced, at no cost to NASA, a new line of wave diagnostics, which were used to probe the wave structure in the plasma. A collaboration with the Plasma Research Group of the Australian National University (ANU) connected the ASPL team with the pioneering work of Dr. Roderick Boswell, co-inventor of the helicon discharge, and his team in Canberra. Early in 2002, Dr. Boswell visited the ASPL accompanied by Mr. Orson Sutherland, one of his graduate students at ANU. Due to unforeseen circumstances, Dr. Boswell had to delay his return to Australia and had a chance to spend a few weeks with the VASIMR® team. During his visit, he provided excellent insights into helicon physics and potential antenna configurations, which the team could try. One of these designs, a saddle-type antenna, was actually incorporated in a magnetic cusp configuration in the laboratory and briefly studied. While cusp magnetic field configurations would not be appropriate for propulsion systems, the experiment generated useful data on the physics of helicons and the characteristic electric fields that accompany these plasma discharges. Other fruitful and synergistic relationships were established with Dr. Shunjiro Shinohara of Kyushu University and Dr. Genta Sato of Fukuoka University in Japan, who were experimenting with novel helicon designs. In the US, Dr. Patrick Colestock and Dr. Max Light of the Los Alamos National Laboratory, with their management's approval, provided access for the VASIMR® team to valuable expertise in RF physics and technology, using their own internal funds.

In 2001, the team turned its attention to achieving direct ion heating in experiments with ICRH. Ion cyclotron heating experiments in the second stage of the engine had been postponed due to limited funds. However, with the relative success of the helicon program, it was time for the team to turn its attention to the RF booster. These initial studies were enabled with the installation of a fourth magnet for field shaping in the ICRH section. Experiments began by operating at the helium fundamental frequency. However, the initial data showed poor antenna "loading," meaning that the RF energy was not being efficiently absorbed by the plasma. Initial observations attributed this behavior to the small diameter of the plasma column produced by the helicon. The team therefore turned its efforts to increasing the plasma size at the helicon, as well as continuing to work on increasing the plasma density with the goal of achieving full ionization of the feedstock gas.

Despite the low coupling of the waves to the plasma with that configuration, these experiments yielded some intriguing measurements that indicated a potential mechanism for the waves to transfer energy to the particles through resonance at the second harmonic frequency. These measurements were reinforced by a new sensitive force sensor, which had been installed in the device to measure the plasma momentum flux. This activity followed earlier work, led by Dr. T. F. Yang and others at the MIT Plasma Fusion Center in 1994 [1]. In the new system, the diagnostic involved the insertion of a small paddle, installed on the end of a sensitive beam of alumina. Small deflection forces imparted by the impinging plasma could be measured through sensitive strain gauges positioned at the fixed support of the beam. The work was part of a PhD thesis research conducted by Mr. Greg Chavers, an employee of NASA Marshall Space Flight Center (MSFC). As part of the developing collaboration with the MSFC, Mr. Chavers had been assigned by his management to work on the VASIMR® engine under the guidance of Dr. Chang Díaz. He successfully completed his PhD thesis at the University of Alabama in Huntsville in 2002.

The specific mechanism for energy coupling at the second harmonic was not understood, but pointed to the potential of as yet unexplored processes, other than pure ICRH, to couple energy to the plasma in the second stage of the engine. These tantalizing possibilities led the VASIMR® team to refer to the second stage as the "RF booster" and not the ICRH stage, so as to not close the door on other potential mechanisms outside of ICRH that could become important later on. This was especially true in regards to accelerating heavier propellants whose gyro-frequencies were outside the range of wave frequencies that could be stimulated with the existing magnetic field strength of the device.

In April of 2001, the ASPL team received the prototype Du Pont high temperature superconducting magnet and began a series of acceptance tests in a small surplus vacuum chamber they had reconditioned for this purpose. A superconducting magnet is needed to provide the strong magnetic field required by the engine, while minimizing power requirements and mass. The ASPL prototype coil was designed to be similar to what would be required for space flight. A less expensive, low temperature superconducting coil of much larger size would eventually be integrated in the experimental apparatus as an important preliminary step in the superconducting technology maturation process. The high temperature superconductor, however, would be the end goal.

The Du Pont magnet used the most advanced high temperature superconducting tape commercially available at the time, denoted as BSCCO-2223 (short for Bismuth Strontium Calcium Copper Oxide), a material that became superconducting below approximately 106 K, and whose ideal operating temperature would be in the range of 40 K. The magnet project was led by one of the leading experts in the field, Dr. Christopher Rey at Du Pont, and was internally funded by the Du Pont Corporation through the intervention of Capt. (USN, Ret) David M. Walker, a former astronaut and a strong supporter of the VASIMR® project. David had followed the research over several years, as he and Dr. Chang Díaz had developed a close friendship as they worked together in astronaut support operations at the Johnson and Kennedy Space Centers. He had retired from NASA and was now working as Global Industry Manager for Aerospace for the DuPont Corporation in Wilmington, Delaware.

By early 2002, the VASIMR® team achieved 39 K, a temperature well within the superconducting range, in a volumetric and thermal facsimile of the superconducting magnet, using a simple low power, off-the-shelf cryocooler. This test paved the way for testing the actual magnet, which occurred in the fall of that year. In August, the team achieved the transition to the superconducting mode in the actual magnet and in September, they initiated powered tests. Measurements of the resulting magnetic field followed the predicted values closely. Sadly, David Walker never got to see the excellent results obtained with the new superconducting magnet prototype he had helped to develop. He passed away after a brief battle with cancer in April of 2001, but his timely support for the VASIMR® team and his contribution to the realization of the technology will never be forgotten.

As the physics of the engine became better understood, the VASIMR® team began to gravitate more towards the engineering of the device and the technological challenges that undoubtedly awaited. One of the key issues the team recognized early on was that of thermal management. The plasma, particularly in the helicon source, was a strong emitter of UV light. This radiation, stemming primarily from the neutral-to-ion transitions of the propellant atoms, represented a thermal load that would end up being deposited as heat on the structure of the rocket, primarily on the first plasma-facing wall of the device. This heat had to be removed before it soaked into the superconducting magnet assembly, only a few centimeters radially away. In 2002, one of the ASPL students from MIT, Ms. Kristy Stokke, conducted a detailed measurement of the thermal environment near the helicon antenna with an array of thermocouples spaced along the helicon tube, as well as light emission measurements. These measurements gave the team a first-hand look at the patterns of heat deposition on the first wall. Her work was later continued by others, such as Daniel Berisford, a graduate student at the University of Texas at Austin, whose PhD thesis explored in detail the thermal environment of the helicon ionizer stage. In addition, an outstanding team of graduate students and faculty from Ireland's University College Dublin, in collaboration with scientists from the newly formed Costa Rica subsidiary of the Ad Astra Rocket Company, further characterized this challenging thermal environment and published a number of peer reviewed papers on their results.

Heat removal from the helicon first wall was complicated by the fact that the wall material was typically a ceramic insulator, and ceramics, in addition to being poor electrical conductors, are also notoriously poor heat conductors. There is one, however, that does not follow this pattern, namely diamond! The prospect of using a diamond helicon tube was intriguing and the team began to explore the technologies of chemical vapor deposition (CVD) manufacturing of crystalline diamond. The idea became more real when Christopher Stott, President of ManSat Inc., and husband of Astronaut Nicole Stott, came by the laboratory to discuss a new type of industrial diamond product his company on The Isle of Man, in the British Isles, had come in contact with. The product could presumably be manufactured in large pieces that could meet the requirement for the plasma source. Some samples were tested at the JSC materials laboratory and examined with an electron microscope. Other exploratory tests, conducted at the ASPL, subjected the material to hours of continuous exposure to a deuterium plasma. Large samples, however, were expensive and other materials were found to provide comparable results. These studies led to major technology advances in addressing the thermal management and materials aspects of the VASIMR® engine and provided the basic path to the technology's present maturity.

Additional work also continued on the evaluation of another potential flight experiment, a flight demonstration of a 25 kilowatt engine on the ISS. The team completed a preliminary design of such a system and a full-scale mockup was built. The theory group of the VASIMR® team also made progress in describing the likely mechanism responsible for plasma detachment from the magnetic nozzle, a subject of growing controversy happily promoted by those who adamantly opposed the project. Plasma detachment, although theoretically predicted, was difficult to experimentally verify in the existing apparatus.

Achieving a good healthy plasma in the ionizer was only half the battle, however, and it was important to demonstrate that the ICRH process could be achieved in the second stage of the device. The goal to demonstrate single pass ion acceleration meant that the tandem mirror configuration of the original MIT experiment had to be modified by eliminating the central cell of the apparatus. The new configuration became much simpler and shorter, consisting of only the four evenly spaced end cell coils, with each independently controlled to achieve the desired axial magnetic field profile. The helicon ionizer was located between the first two coils and the second stage ion cyclotron RF booster was installed further downstream.

The experimental results of the VX-10 in the early 2000s had been excellent and gave the team strong confidence in the physics of the device. Two important achievements occurred during this early period: First, the efficiency of the helicon ionization process was demonstrated in the axial magnetic asymmetry and rapidly flowing plasma characteristic of the VASIMR® engine; and second, with proper optimization of the controlling parameters, complete propellant burn up was achieved, ensuring the absence of neutral particles in the second stage. Today, efficient plasma production at 10^{20} particles/m^3 is responsible for the engine's inherent high power density.

In early 2000, enthusiasm for the project in the plasma physics community had become contagious and despite the lack of formal institutional acceptance, the research team grew to a unique amalgam of numerous scientists and engineers with diverse but complementary skills, as well as funding sources. Various small grants, obtained at different times, supported graduate students and professors at universities such as the University of Texas, Rice University, the University of Maryland, Houston, Michigan, Alabama, MIT and others. In Houston, the Lockheed Martin Company deployed Dr. Andrew Ilin, a mathematician from Russia and recent employee, and supported him for a year on company funds, though these were later refunded to Lockheed Martin by decision of the JSC Management. The Safety Reliability and Quality Assurance Division of the Johnson Space Center also assigned Mr. D. Scott Winter to the ASPL, while the JSC Engineering Directorate assigned Mr. Andy Petro, a systems engineer working in the Office of Exploration. NASA MSFC assigned one of its employees, Mr. Greg Chavers, to the ASPL in support of his PhD research, as well as Carol Dexter, a mechanical engineer, who supported the ASPL for one summer and built a high voltage enclosure for one of the impedance matching circuits of the apparatus. Other engineers from JSC and MSFC came to work at various times on the project and a number of external research groups also became involved, leveraging some of their own funding to support the project, including world-renowned investigators from the National Laboratories at Princeton, Los Alamos and Oak Ridge, as well as international groups from Sweden, Russia, Japan and Australia. The relationship with the High Technology Center in Costa Rica lasted more than four years, up to the formation of Ad Astra Rocket Company, bringing eight graduate level researchers from Costa Rica for internships in the laboratory. This expanded the research to participants in the developing world, a personal goal of Dr. Chang Díaz since his early years at NASA.

Another major change in the operation of the laboratory came in late 1999, when Dr. Chang Díaz obtained permission from his management to bring the ASPL team under a single contractor. A small start-up company by the name of Muñiz Engineering Inc., had qualified for government funding as a small business and was looking for a stable engineering services contract. Its leader, Mr. Edelmiro Muñiz, was a former US Air Force officer and someone whom Dr. Chang Díaz had gotten to know and admire over the years as a man of integrity and honesty and as a hard working business entrepreneur. Earlier that year, Dr. Chang Díaz had had the honor to present Mr. Muñiz with a business award and recognition from the Houston Hispanic business community.

At a meeting with Dr. Chang Díaz at the ASPL, Mr. Muñiz briefed him on the capabilities of his company and proposed the idea of hiring the small ASPL team of individual contractors and deploying them as an integrated workforce under his company. Dr. Chang Díaz became immediately interested in this proposal and wasted no time in presenting it to JSC Center Director Abbey and some of his key officers. While technically such a move would not introduce added value to the project, programmatically it would. The integration of an approved NASA small business contractor was strategic, as the small company would certainly represent an important ally in the continuous struggle for survival, adding much needed programmatic legitimacy – and hence, stability – to the laboratory. Moreover, it was a win-win proposition, in which the ASPL personnel would benefit from a better structure of employment benefits and the small company would benefit from the NASA funding and the prestige of such an advanced technology project. But in the end, it all came down to one basic question: where would the funding come from?

5.5 The VX-10 configuration discarded the central cell, resulting in a much simpler design.

Mr. Randall Gish, Head of Procurement at NASA JSC, provided the answer to that question. Randy was a visionary leader within the JSC management and had come to appreciate the importance of the VASIMR® project at JSC. He understood the unusual politics in which the project had evolved and now existed, but also recognized the strong technical progress the ASPL team had achieved under such adverse conditions. He was not inclined to see it die. After several meetings and presentations, he agreed to award a small, sole-source contract, subject to annual review, to Muñiz Engineering to hire the ASPL team and provide the integrated support services to the project. Mr. Thomas Krenek transitioned from the Yang Technologies contract as the NASA Contracting Officer overseeing the new contract with Muñiz, while Dr. Chang Díaz continued in his role as Contracting Officer Technical Representative (COTR) and ASPL Director. Six employees were hired through the new contract: Dr. Squire, who transitioned from Yang Technologies, Dr. Andrew Ilin and Mr. Greg McCaskill (both Lockheed Martin employees), Mr. Garland "Buddy" Goebel (a technician with McDonnell Douglas), Mr. Verlin Jacobson, a recent MS level graduate from the University of Texas at Austin, and Ms. Sandra López, who became the team's Administrative Assistant.

THE GATHERING STORM

As the world welcomed the 21st Century, the VASIMR® project continued to enjoy the attention of high-level NASA officials, including Administrator Daniel Goldin, who had been a supporter since his first visit to the laboratory on January 18, 1996. Funding for the project, while still erratic and uncertain, had markedly increased through an aggregate of sources, topping over $1.1 million in 2000. At JSC, the plasma rocket had become a "poster boy" project and a prominently featured part of the Center's VIP tour and public relations events. The increased funding resulted in more rapid progress in the research, with major physics questions being answered through carefully designed experiments. Following the prestigious Rotary Stellar Award in 2000, the VASIMR® team continued to be recognized by the US space community and, in 2001, Dr. Chang Díaz was awarded the Wyld Propulsion Award by the American Institute of Aeronautics and Astronautics (AIAA). The award is given for outstanding achievement in the development or application of rocket propulsion systems.

Yet, despite the diverse and seemingly growing pockets of support, the ASPL still lacked a permanent home. Institutional advocacy for the project, although growing, was widely scattered and the laboratory remained under the uneasy foster care of the Astronaut Office. Inside JSC, the ASPL team sought to develop a working relationship with its closest logical partner, the JSC Office of Exploration, a separate team whose charter included the development of the official NASA mission architecture concept for human Mars exploration. The group was part of the JSC Engineering Directorate and had survived, under various names, through multiple organizational changes. This relationship, however, was a difficult one and never fully crystalized, as the group did not consider the VASIMR® engine to be a credible human exploration technology and a nuclear-electric approach to Mars exploration was not in the architectural path they had chosen to follow. Instead, a conventional, all-chemical propulsion architecture had been selected, with an advanced option based on the 1960s NERVA nuclear-thermal rocket.

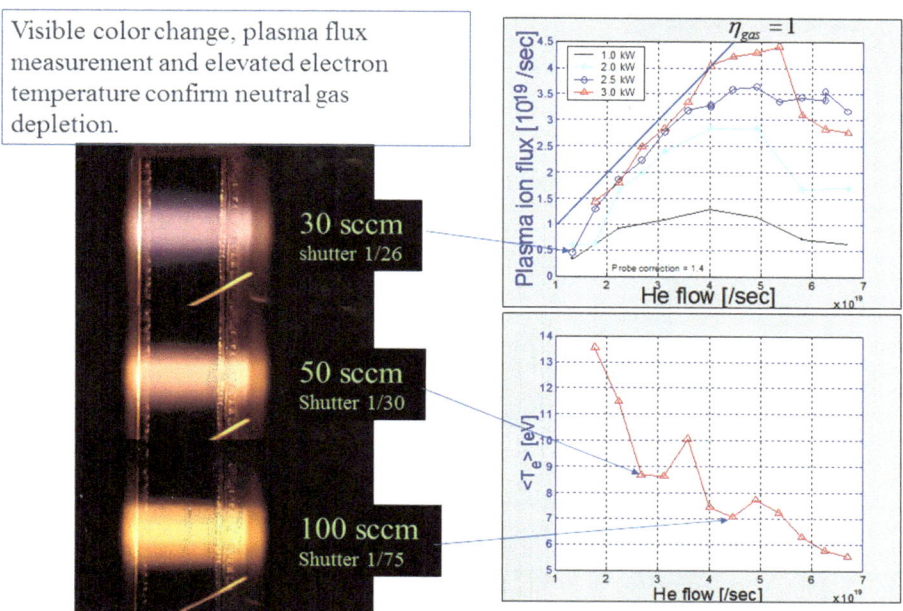

Visible color change, plasma flux measurement and elevated electron temperature confirm neutral gas depletion.

5.6 Visible discharge color changes are associated with total propellant ionization. Also, plasma flux vs. neutral input flow for helium (upper right) shows the "one for one" (neutrals for ions) correspondence associated with total propellant utilization (burn up). The elevated electron temperature at low mass flow rates also confirms neutral gas depletion.

There was an apparent inconsistency in NASA's nuclear posture. While the nuclear-thermal propulsion (NTP) approach appeared to enjoy institutional acceptance, the high power nuclear-electric propulsion (NEP) option did not. The all-chemical and nuclear-thermal architectures produced missions with long transit times and were heavy on the propellant load. The high power nuclear-electric approach, on the other hand, offered potentially drastic reductions in transit time and provided greater operational flexibility, wider launch windows and more attractive abort options in the event of major malfunctions en route. It was clear, however, that the NEP approach was less technologically mature, but even so, given the acceptance of a nuclear option, the greater advantages of NEP provided a powerful incentive for its serious pursuit and the VASIMR® team felt that it deserved due consideration in the architectural study.

Nuclear-electric power had indeed been considered in a couple of low power propulsion applications, such as the SP-100 project in the 1980s and Project Prometheus in the 1990s. Both of these were relatively short-lived and never made it into space. Despite its critical importance, US space planners had continued to avoid the propulsion applications of multi-megawatt nuclear-electric power, focusing instead on low power systems at the 100 kW level; better suited for surface power applications but whose advantages, as compared to increasingly powerful solar-electric technologies, were difficult to discern. This condition persists to this day. It does not help that, traditionally, US expertise in nuclear-electric power does not reside at NASA but exists instead within the DoE. In the absence of a higher mandate, this

5.7 The ASPL Team (circa 1999): From L to R (Standing) – Andrew Ilin, Tim Glover, Jared Squire, Franklin Chang Díaz, Garland "Buddy" Goebel, Greg McCaskill, D. Scott Winter. Foreground – Jeff George, Tri Nguyen, Tim Graves, Kristy Stokke, Carol Dexter, Andrew Petro

institutional decoupling of power and propulsion could be a significant inhibitor in NEP's ability to become a front stage program. This is not the case with NTP, which is fundamentally a heat transfer and materials problem and fits more easily within the NASA purview.

More headwinds were on the way, as major changes were taking place at NASA which, once again, were sure to disrupt the VASIMR® engine project. In February of 2001, JSC Center Director, Mr. George W. S. Abbey, resigned amid budgetary and political issues. Administrator Goldin appointed Mr. Roy Estess, Director of NASA's Stennis Space Center in Mississippi, as his acting successor, but in late 2001, Goldin himself left the agency and was replaced by Sean O'Keefe, a financial administrator from the Office of Management and Budget. Deep changes in the organization were being implemented and soon after, rumors began to circulate of the potential shutdown of the ASPL and its VASIMR® engine project at JSC. The team had endured these headwinds before, however. They had concluded that continued technical progress, combined with a strong communication effort of its results, was the best way to combat those who wished to see the project shut down. In order to educate the new JSC leadership, Dr. Chang Díaz arranged for a short briefing on the project for Mr. Estess. The meeting was organized by Capt. John W. Young, Dr. Chang Díaz's close friend and longtime VASIMR® supporter.

5.8 Dr. Jared P. Squire receives the Rotary Stellar Award on behalf of the ASPL team from astronaut Eileen Collins at the 2000 Rotary Gala dinner, held at Space Center, Houston.

Dr. Chang Díaz was also assigned to fly on STS-111 in 2001, his seventh space mission. This would be a 14-day logistics and Space Station assembly flight on the Space Shuttle *Endeavour*, with three planned EVAs (Extra Vehicular Activities, or "spacewalks") which Dr. Chang Díaz would lead. Underwater training at the Neutral Buoyancy Laboratory (NBL) would be intense over many months leading to the flight. Fortunately, the NBL happened to be in the same facility as the ASPL and this close proximity made it easy for him to carry out his underwater training with minimal impact to his director duties at the ASPL. Moreover, Dr. Chang Díaz had assigned Dr. Squire to lead the laboratory and the team in his absence. Dr. Squire had proved to be an outstanding researcher; he was calm, methodical and organized, and as had been the case on his previous space missions, Dr. Chang Díaz had full confidence in his ability to lead the project in his absence. However, as his training progressed, the institutional headwinds began to reach gale force, as the rumors of the laboratory shutdown became more frequent, peaking as Dr. Chang Díaz and his crew entered quarantine just one week before flight. It was impossible to not suspect a pre-meditated motive for such timing.

The established electric propulsion community continued to staunchly dismiss the VASIMR® engine as a credible electric propulsion technology. Instead, they had paced the evolution of high power EP in terms of what traditional ion engines and Hall thrusters could hope to achieve. While the VASIMR® team was already discussing thrusters inherently capable of hundreds of kW, ion engines and Hall thrusters remained at power levels in the single digits. A great deal of money and time had been invested in those technologies, however, and the established groups were not receptive to change – even less so to welcome a potential challenge in the high power arena where the future was known to lie. As the ASPL team grew and the VASIMR® engine project became more widely known, obscure pressures for NASA to discontinue the research continued to build up. As had been done once before, the procedure to accomplish a project shutdown would be by means of a "peer review."

As the launch date for STS-111 approached, Dr. Chang Díaz became increasingly concerned about the fate of his team and that of the ASPL while he was in space. There appeared to have been some truth to the rumors, as he learned indirectly of a NASA plan to conduct a peer review on the VASIMR® project. It was a familiar tune that brought back memories for him of a decade earlier, when the fledgling project at the MIT-PSFC had fought for survival in front of a hastily assembled review panel. That panel had convened as a virtual surprise, giving the researchers little time to prepare. From Dr. Yang and Dr. Chang Díaz's perspective, its goal was to simply impart some semblance of legitimacy to the arbitrary shutdown of the project. This time, once again, the unknown enemy's footsteps were being felt, and, curiously enough, they appeared to coincide with Dr. Chang Díaz's impending mission into space.

His fears for the fate of his team and the laboratory led him to approach Capt. Young, who listened to his concerns and took immediate action. He had been appointed Associate Director, Technical, responsible for the technical, operational and safety oversight of all Agency Programs and activities assigned to the Johnson Space Center (JSC). There was little else Dr. Chang Díaz could do now to protect his project, as he entered into quarantine in late May of 2002, a week prior to launching into space. Fortunately, Capt. Young intervened in a vehement and timely fashion on behalf of the team and the project. On the eve of Dr. Chang Díaz's launch, Capt. Young reassured him that his project would be safe while he was away. The rumors about an impending peer review of the VASIMR® were indeed real and the plan appeared to have been concocted somewhere within the intricacies of the space agency's bureaucracy. However, Dr. Chang Díaz's focus was now on the success of STS-111 and he flew into space with the knowledge that the review would wait until his return to Earth. Unbeknown to him, however, on Wednesday, June 5, 2002, the day he launched into space, the ASPL team received notices from NASA that their contract with MEI was cancelled and that they were therefore to be locked out of the laboratory.

Thus, an extraordinary situation began to brew on Earth for the VASIMR® project while Dr. Chang Díaz carried out his extra vehicular tasks in space on the construction of the ISS and the repair of the Canadian robotic arm. With the ASPL team effectively disbanded and locked out of the laboratory, it appeared as though the stage was being set for a peer review, not of the complete project, but solely of Dr. Chang Díaz himself, without the support of his team or of an active laboratory. Under such unfair conditions, it would be extremely difficult to survive. Fortunately, and also without his knowledge, the news of the ASPL team's contract cancellation had reached the office of Senator Bill Nelson of Florida, a former crewmate and friend of Dr. Chang Díaz, who had followed the development of the VASIMR® project over the years and was quite familiar with the high caliber of the research. He immediately made inquiries with Administrator O'Keefe and requested an update on the project. It was not clear to Dr. Chang Díaz, upon his return to Earth, how far in the chain of command the project-crippling attempt had reached in his absence.

However, Senator Nelson's timely call seemed to have stopped the process and provided a reprieve on the cancellation of the MEI contract. The team could return to the laboratory and prepare for the peer review.

Meanwhile, more management changes were taking place at JSC. A new Center Director, a military man by the name of Jefferson "Beak" Howell, who was a US Marine General and former fighter pilot, had been appointed in the spring of 2002 by Administrator O'Keefe to succeed Roy Estess. Dr. Chang Díaz came to know him on the eve of his STS-111 launch. Gen. Howell was a pleasant, no nonsense individual who focused on "the big picture." He had become aware of the ASPL research and the planned peer review and had assigned his Deputy Director, Mr. Randy Stone, to oversee the process. Randy was the right man for the job. An experienced systems engineer and former flight director, Mr. Stone was also quite aware of the internal politics of the agency and the unusual journey of the VASIMR® project from MIT to the JSC in Houston. Upon Dr. Chang Díaz's return to Earth, Mr. Stone played an important role in securing both the proper funding to ensure a fair and unbiased review and that sufficient time would be allocated to the ASPL team and its collaborators to prepare for it.

5.9 STS-111, June 2002: "Hanging in the breeze." Dr. Chang Díaz, standing on the end of the ISS robotic arm (Canadarm-II), performs the first of three extravehicular activities (EVAs). On his seventh mission into space, he and his crew contributed to the assembly and development of the International Space Station. On this mission, Space Shuttle *Endeavour* also ferried the ISS Expedition 5 crew, Valery Korzun, Dr. Peggy Whitson and Sergei Treschev to the station, and returned the ISS Expedition 4 crew, Yuri Onufriyenko, Daniel Bursch and Carl Walz back to Earth.

By June 10, 2002, while Dr. Chang Díaz was in space, preparations for the peer review had already commenced on the NASA side. The process appeared to originate from the Office of the Assistant Associate Administrator for Advanced Systems, headed by Mr. Gary Martin who later that year was named NASA's Space Architect. Dr. Chang Díaz had met Mr. Martin and had been impressed with his long-range vision approach and strategic thinking. While he remained alert to any signs of foul play, Dr. Chang Díaz was hopeful that his team and project would get a fair review. The process organization that resulted from the Washington directive was odd and awkward, staying clear, for some reason, of the Astronaut Office, Dr. Chang Díaz's direct management chain of command. Instead, the information seemed to flow to the ASPL through a rather circuitous route; first through Mr. Les Johnson, Manager, In-Space Transportation Investment Area at the MSFC in Huntsville, Alabama, and then through Mr. Douglas R. Cooke, Chief of the JSC Office of Exploration.

Dr. Chang Díaz and the STS-111 crew of *Endeavour*, which also included Commander Kenneth Cockrell, Pilot Paul Lockhart and French astronaut Phillippe Perrin, returned to Earth on June 19, 2002, after 14 days in space. The mission was a complete success, including a major repair to one of the active joints of the Canadian robotic arm, which had failed a few months before and needed to be replaced. The repair was accomplished in a 7-hour spacewalk, one of three conducted by Dr. Chang Díaz and fellow crewmate, Phillippe Perrin. The *Endeavour* also ferried ISS Commander Valery Korzun and Flight Engineers Peggy Whitson and Sergei Treschev, designated as the ISS Expedition 5 crew, and brought back to Earth ISS Commander Yuri Onufriyenko and Flight Engineers Daniel Bursch and Carl Walz, the ISS Expedition 4 crew, after their 182-day residence aboard the station.

With seven flights under his belt, Dr. Chang Díaz had become well adapted and accustomed to the physiological changes brought about by flying in space. By today's standards, his flights, although numerous, had been relatively short, but this also meant that he could recover rapidly from each spaceflight. Following the standard management and operations debriefs and other post-flight protocols for STS-111, he returned his focus to the ASPL and the VASIMR® engine project. It was time to prepare for the expected peer review.

On July 11, 2002, the first formal notification of the impending review arrived in an e-mail communication to Dr. Chang Díaz from Mr. Cooke. The message contained the first description of the review planning process, which had started in early June with a proposed review date of August 14, 2002. At first glance, it appeared to Dr. Chang Díaz and his team to be an extremely ambitious schedule to assemble a well-balanced group of experts with undoubtedly busy agendas. Nonetheless, the ASPL team began immediately to prepare for what was shaping up to be the most important test for the project so far.

By early August, their predictions had proven correct, as it was clear that the mid-month target date planned by NASA was unrealistic and unattainable. As soon as he returned from space, Dr. Chang Díaz had discussed the impending review with, and sought the advice of, his close friend Professor Samuel C. C. Ting, leader of the AMS project. In their discussions, Professor Ting advised caution on understanding the composition of the review panel. He had seen – and been the subject of – many review panels in his career, some of which had been designed with the sole purpose of killing the project. He warned Dr. Chang Díaz to be suspicious of the motives of the review panel. He indicated that, while Dr. Chang Díaz could not choose the panel members, he had the right to veto any reviewer whom he felt would be unduly biased or had a conflict of interest. Dr. Chang

Díaz took this advice to heart and began a careful study of the recommendations for potential reviewers that arrived from Mr. Johnson at MSFC.

The intense negotiations took time. Some of the proposed reviewers were known VASIMR® detractors with little knowledge of the underlying physics of the engine. Others were known scientists from the traditional electric propulsion community, but whom the VASIMR® team considered would be fair and unbiased. The VASIMR® engine was an unusual system and the review panel's composition had to be equally unusual. It had to reflect the culture convergence that Dr. Chang Díaz had sought to achieve for more than two decades. It was imperative to achieve a good technical balance between unbiased experts from the traditional electric propulsion community and those well versed in areas unique to VASIMR®, such as superconducting technology, RF power and high density plasma physics. Many of the latter originated predominantly from the field of magnetic fusion research and departed significantly from traditional electric propulsion. Preparations for the review continued through August and a new review date was set for October 8, 2002.

While the review date had been set, it was apparent that the MSFC organizers were not able to fully anticipate the level of effort that would be required to assemble a qualified VASIMR® peer review panel in a timely manner. As the days and weeks passed, some of the potential reviewers complained about the lack of organization and others, citing frustration with NASA's lack of process coordination, simply withdrew their candidacies. Nonetheless, Dr. Chang Díaz had communicated the October 8 date to the extended VASIMR® team and made plans to bring as many of the VASIMR® scientists as he could to Houston to support the team in the review. Several of them made plans to arrive in Houston a couple of days earlier to take part in an informal review preparation workshop and share last minute findings in each of their particular development areas.

On October 8, 2002, more than 30 VASIMR® scientists convened in Houston and joined the ASPL team to support their colleagues during the peer review, some traveling at their own expense from as far away as Sweden, Australia and Italy. The gathering had been timed by the ASPL team to provide the review panel with comprehensive access to the project and bear witness to the strong scientific collaboration that the VASIMR® project had established with US universities, National Laboratories and international research institutions. Unfortunately, from the reviewing team, only Mr. Richard Fischer, an observer from NASA Headquarters, managed to attend on that day. He had arrived to represent Mr. Gary Martin, the review sponsor from NASA in Washington DC and was, as with the rest of the VASIMR® team, unaware that the review organizers at MSFC had decided, at the last minute, to postpone the review date one more time. They had continued to run into additional organizational and logistical glitches, which prevented the panel members from securing a quorum in Houston on the appointed date.

It was too late to call the gathering off. The VASIMR® team was already convened at the ASPL. It was an extremely disappointing situation, as Dr. Chang Díaz had committed the project's remaining travel funds to bring the team to Houston and there was no more funding available for a second chance. But there was nothing to do, except to make use of the opportunity to work together to update the team on the collective progress. Therefore, on October 8, 2002, Dr. Chang Díaz opted to hold a two-day workshop with the entire team to thoroughly address the questions that had been submitted by the peer review process.

It was an unfortunate turn of events, which was further complicated by the expenditure of the remaining ASPL travel funds. As the review organizers searched for another schedule date, the VASIMR® team did not have the means to bring everyone back to Houston a second time. Fortunately, Mr. Randy Stone assured Dr. Chang Díaz that the required travel funds would be made available to bring back the extended team on the review's new scheduled date. There was some level of comfort in that promise. However, beyond the immediate issue of the travel funds, another problem loomed more ominously. The end of the year was approaching and the ASPL contract was due to expire at the end of November. At that point, pending the outcome of the review, there were no more sustaining operational funds identified for the ASPL.

On October 15, 2002, Ms. Alice J. Purcell from the JSC Office of Procurement contacted Dr. Chang Díaz inquiring about his intention to exercise the option for a six-month extension of the Muñiz Engineering contract, which was due to expire on November 30. While there were no funds identified for the laboratory past that date, it was important to anticipate the possibility that some would appear in the intervening weeks and have the paperwork ready just in case. After all, funding uncertainty had been a way of life for the ASPL team since its inception. It was clear to Dr. Chang Díaz that the new peer review date would fall sometime in late November and, assuming a successful review, it was likely that funding would be immediately allocated to keep the project going. If so, a rapid-fire action would be required to prevent voiding the contract. Fortunately, Ms. Camille Goodwin, one of JSC's most capable Budget Integrators, Ms. Martha Bishop,

5.10 The extended VASIMR® research team gathered in Houston on October 8, 2002 to support the peer review. It included investigators from the Los Alamos and Oak Ridge National Laboratories, the Australian National University, the Alfvén Laboratory in Sweden and scientists from several US universities, NASA facilities and private industry. Mr. Richard Fischer (7th standing from right), an observer from NASA Headquarters, represented the NASA Exploration Team, the review charter organization.

Business Manager for Flight Crew Operations at JSC, and Victoria Osteen from Procurement, were big supporters of the ASPL and always ready to help. Their timely assistance – and that of Ms. Purcell – always made the bureaucratic gears move at lightning speed once the management decisions had been made.

THE VASIMR® PEER REVIEW

The VASIMR® peer review finally convened at JSC on November 20, 2002. It was a grueling examination for the project, but the team was thoroughly prepared and did an outstanding job in answering all the questions. In the intervening weeks since mid-July, an important physics breakthrough had been achieved: the complete ionization of the injected propellant. The achievement came in October, with the installation of a modified helicon tube. Repeatable measurements with the modified design indicated complete gas burnup (100 percent of the input propellant turned to plasma). This result produced an important side benefit, given the pumping limitations of the laboratory, as the vacuum chamber was able to reach much lower pressures during the discharges. The reduction in the neutral population also reduced the charge exchange and radiation power losses produced by the presence of a high pressure background gas, thus making the system inherently more efficient.

The complete ionization of the propellant was an effect that had been theoretically predicted but had not been achieved experimentally until a new helicon antenna was introduced, along with the modified discharge tube. The experimental data showed that all the propellant injected was actually efficiently ionized. The team had been working with both helium and deuterium but other gases were expected to work as well. It was a great achievement, just in time to answer one of the pressing questions being asked by the review panel: Will the VASIMR® first stage be efficient in ionizing the propellant? The answer was proven to be a resounding yes!

The peer review panel was led by Mr. Gordon Woodcock, an experienced aeronautical engineer who had retired from the Boeing Company but remained active in the aerospace field as a private sector consultant with Grey Research and also worked under contract with the MSFC. The other panel members included Dr. Roald Sagdeev, a world-renowned plasma physicist originally from Russia, who had emigrated to the United States during the breakup of the Soviet Union and was now a distinguished professor of plasma physics at the University of Maryland. He was joined by Dr. Joseph Minervini, an expert on superconductivity from MIT and Professor Miklos Porkolab, an expert in RF heating of plasmas and Director of the MIT Plasma Fusion Center. Adding expertise in high density plasma physics was Dr. Samuel Cohen from the Princeton Plasma Physics National Laboratory. From the more traditional electric propulsion community, NASA included Dr. Edgar Choueiri, who had succeeded Professor Robert Jahn as Director of Princeton University's Electric Propulsion and Plasma Dynamics Laboratory; Dr. Mark Cappelli, an expert on Hall Effect Thrusters from Stanford University; Dr. Ronald Cohen, Director of Propulsion Science at the Aerospace Corporation; and Dr. Mike Gruntman, a Professor of Astronautics at the University of Southern California.

The VASIMR® team's preparations prior to the actual review had been intense and thorough. To many of the team members, the review process was reminiscent of their own PhD qualifying exams and theses defense of years past during their student careers. There was a written exam, submitted to the VASIMR® team by the panel a couple of weeks before the review. The VASIMR® team submitted its written responses on November 14. These gave the reviewers time to digest the information prior to a full day of oral presentations, which were scheduled for Thursday November 22. Two NASA appointed rapporteurs, Dr. Lee Morin and Mr. Jaime Forero, witnessed the presentations and discussions. After the oral presentations and questions, the panel deliberated in private on Thursday evening and presented its conclusions and recommendations to the VASIMR® team on Friday November 23.

The peer review panel was extraordinarily thorough and probing and the VASIMR® team was equally meticulous and comprehensive in its responses to the panel's questions. These questions addressed physics, engineering, technology and diagnostics, as well as trajectory studies and mission design. The panel also inquired about the makeup of the VASIMR® team and its management and organization. There were further questions regarding the output and caliber of the team's publication and conference proceedings and the panel was keen on knowing when more peer reviewed publications were expected to appear, an interest that could be reconciled with the academic culture of the reviewers (there were no reviewers from traditional private industry). On this last point, the VASIMR® team made it very clear that it already had sufficient material and interesting results for a large number of articles in refereed publications and that it expected to be submitting a number of these papers in the following months.

At no time did the panel address the central, albeit non-technical issue in VASIMR® research, namely the erratic and unpredictable history of funding for the project and its lack of a stable programmatic home within NASA. For more than 20 years, the project had been an institutional orphan; yet the grass-roots growth from a diverse community of first class scientists was indisputable and this fact was evident during the panel's oral examinations in Houston, where most of the VASIMR® team had once again convened. The reviewers' subsequent visit to the ASPL laboratory presented them with a first class facility, which was clean, safe and well organized. One of the chartered questions posed by the panel inquired about the potential of the VASIMR® engine to open new and unique operational capabilities that other concepts may not. This was a central question that was answered in the affirmative with a great deal of supporting data.

The issue of competing approaches to VASIMR® electric propulsion also arose and the team endeavored to point out the differences and the advantages of the VASIMR® system to outperform those competing approaches. For example, among competing electric thrusters such as ion engines, Pulsed Inductive Thrusters, Hall Effect Thrusters, and MPDs, the VASIMR® concept offered the advantage of having no electrodes, eliminating a major limiting factor for engine lifetime. VASIMR® also offered a greater range of propellant options including inexpensive, plentiful and non-contaminating gases such as hydrogen and helium, as well as the propellants typically used by the other systems. The potential to operate at very high power levels while retaining a low specific mass compared to ion or Hall thrusters was also highlighted.

The team further described the engine's ability to vary the specific impulse and thrust continuously over a wide range with a single engine and the advantages such a capability provides to the mission designer. For example, a VASIMR® propelled spacecraft can depart or enter the gravity field of a planet with higher thrust and then gradually shift to higher specific impulse for greater fuel economy during the heliocentric transfer. While providing fast interplanetary transfers, this variability could also enable mission abort options which may not be possible with other existing or proposed systems.

The review panel also briefly ventured into more exotic interplanetary propulsion schemes such as solar sails and plasma bubbles. The latter concept had captured the attention of the advanced propulsion community with a proposed magnetized plasma bubble that could "inflate" a magnetic field to several times its empty volume, thereby producing a plasma sail that could utilize the solar wind for propulsion. With these questions, it was evident to the VASIMR® team that some of the reviewers were endeavoring to lump the VASIMR® concept in with the more exotic and futuristic systems, a surefire way of preventing it from going mainstream and having access to realistic budgets. More than a decade after the peer review, this futuristic baggage has been one of the most difficult to shed.

The VASIMR® team nonetheless wasted no time in pointing out the near-term applications of the technology, as solar-powered spacecraft for commercial applications near Earth. However, one of the key terms of reference for the review panel's evaluation of the VASIMR® engine was as primary propulsion for robust, high-powered, nuclear-electric spacecraft designed for human and large cargo missions to Mars and beyond. Therefore, most of the presentations to the committee addressed the capability of VASIMR® spacecraft to reach their destinations without the need for gravity assists or aerobraking, their ability to execute abort maneuvers and real-time reconfiguration to survive a large variety of unforeseen contingency scenarios.

Addressing the limitations of the technology, the VASIMR® team acknowledged the still low system technology readiness level (TRL), as compared with lower power ion and Hall thrusters, which were considered mainstream. The team recognized that both solar- and nuclear-electric space power technologies were still not fully developed. Looking back to 2002, such was indeed the case. However, more than a decade later, the technology of solar-electric power has matured to the point where solar arrays of hundreds of kW are now possible and NASA has begun to speak of high power solar-electric propulsion in a power niche where VASIMR® becomes highly competitive. It was clear to the VASIMR® team in the early days that the engine did not down-scale well to low power. The technology becomes concept attractive at high power, greater than 50 kW. We know today that, for power systems less than 50 kW, the system is not well suited due to many factors, including the reductions in the plasma's useful diameter which renders the magnet system unduly heavy and expensive. Therefore, the VASIMR® engine does not intend to compete in that power space.

The reviewers wanted to know what the main unresolved systems questions were of the VASIMR® concept and its implementation, as well as which issues were endemic to electric propulsion systems in general and which were VASIMR®-specific. At the time, there were – and still are – two major questions endemic to all electric propulsion systems operating at high power. Firstly, there is the performance, mass and configuration of the electrical power source. The nuclear power system and the heat rejection technology that will be available for future engines comprised a particular case. Secondly, all propulsion systems, electric and otherwise, need to be concerned with the availability and storage of propellants. At this point, the team was allowed to dream a little and look far into the future. They looked at VASIMR® as a precursor to fusion rockets. An excerpt from the presentation to the review panel read as follows:

"Looking far into the future we may be allowed to dream a bit. If this is the case, we could envision the possibility of a reversed-field configuration nested within the central cell of a VASIMR®-like geometry. If aneutronic fusion ignition could be achieved in such a plasma, the engine's demand on electrical power (and the fission reactor) would be virtually eliminated, producing a fusion rocket with awesome performance."

To be sure, one of the panel reviewers, Dr. Samuel Cohen, from the Princeton Plasma Physics Laboratory, had studied the potential to generate a reversed-field configuration in a VASIMR®-like geometry. Such structures could, in principle, lead to sufficient particle and energy confinement to trigger a sustainable fusion reaction. The fusion of hydrogen with boron 11 was considered ideal, as its reaction resulted in charged particles only and produced no neutrons, unlike the more popular deuterium-tritium reaction. The charged particles, therefore, would remain confined in the plasma and contribute to its thermonuclear ignition. This was indeed an exciting proposition that, at present, still remains technologically distant.

The review panel required the VASIMR® concept to be designed to address an anticipated NASA programmatic need for human nuclear-electric propulsion missions. This program would enable fast transits to reduce crew exposure to harm; allow demanding missions to be performed for reduced launch mass; exhibit robust operation and high reliability over the design lifetime; provide enhanced abort options for a variety of scenarios over broad segments of the mission; enhance mission flexibility through widened departure windows and provide a power-rich environment for crew subsystems.

These needs translated into specific "near-term" propulsion requirements that were specified in the terms of reference for the review, namely: total system power of 6 MWe; thruster unit power of 1 MWe; specific impulse of 4000 to 7000 sec; efficiency of greater than 50 percent; thruster and power processing alpha of less than 1 kg/kWe and an operating life greater than 600 days. For the "farther term," the total system and thruster power were increased to 20 MWe and 5 MWe respectively, the specific impulse was increased to 10,000 sec, and the engine lifetime was nearly doubled.

The reviewers inquired about mission scenarios and trajectories that could be accomplished with typical VASIMR® parameters, and the mass estimates in kg/kWe that could be generated, at a "reasonable" power level, for the main elements of the VASIMR® system. In addition, the panel wanted to know if there was any experimental evidence to justify the assumption of constant efficiency with varying I_{sp} adopted in the VASIMR® mission studies.

In 2002, it was difficult for the VASIMR® team to assure the fulfillment of all these requirements; however, more than one decade later, the experimental data shows that the system is capable of meeting and exceeding these benchmarks. Given what is known today, the constant efficiency assumption for the mission studies was not far off, although it would be necessary to invoke different propellants for different phases of the mission. In 2002, the VASIMR® team predicted that the engine efficiency, as a function of specific impulse, was dominated by the ionizer stage at low I_{sp} and by the RF booster at high I_{sp}. It was also expected that different propellants would operate efficiently over different I_{sp} ranges. From the results obtained to date, these predictions were remarkably accurate.

The review panel was clearly concerned about efficiency, as some cursory early assessments had erroneously concluded that the ionization cost for the plasma would be too high. Such estimations were based on traditional ionization techniques involving more common capacitive or inductive energy coupling schemes. The efficiency of the helicon discharge to operate in certain regimes of interest to the VASIMR® team changed these estimations. The review panel was thoroughly briefed on the process used by the VASIMR® team to estimate overall system efficiency by accounting for all the known power flow and coupling transitions leading to energy being deposited into the plasma. At the time of the review, the ionization cost was approximately 300eV/ion; however, soon after the review was completed this value was reduced to 200 eV/ion and eventually to less than 100 eV/ion with proper optimization of the source.

The kinetic energy leaving the ionizer and entering the booster was measured by a retarding potential analyzer at about 50-60 eV/ion and, just prior to the review, the VASIMR® team accomplished the major breakthrough of total propellant ionization. The incomplete ionization of the feedstock gas was one of the voiced concerns of the reviewers. However, with a new helicon source and more power at the first stage, the team measured 100 percent of the incoming gas coming out as plasma.

Questions turned to the remaining major experimental unknown; the efficient absorption of energy in the booster section through the process of ion cyclotron resonance heating (ICRH). The team explained that these experiments were still pending in the ASPL and would probably have to await the required funding to extend the laboratory support for at least another six months. The general concept for VASIMR® ICRH still employed the partial trapping of the plasma ions axially between the two magnetic mirrors. However, new studies had proposed that the ICRH coupling was fast enough for the ions to absorb the energy in a single pass through the length of the device on their way out of the nozzle.

The VASIMR® team presented their single pass theory to the review panel, with the understanding that if the single pass approach did not produce good experimental results, the team would revert back to the original mirror trapping concept. The theoretical description of the process was published in the peer reviewed journal, *The Physics of Plasmas*, by two of the VASIMR® team members, Dr. Boris Breizman and his graduate student Mr. Alexei Arefiev from the Institute for Fusion Studies of the University of Texas at Austin [2]. From these investigations, it was estimated that the RF booster would add up to 500 eV to the ions prior to ejection by the magnetic nozzle. Moreover, it was also expected that the so-called "ambipolar[1]" electric field, established naturally by the plasma to maintain "quasi-neutrality" in the plume, would make that number even larger. A year after the review was concluded, the VASIMR® team demonstrated experimentally the single pass ion acceleration and the present VASIMR® system has measured ion energies leaving the booster comparable to those predicted in 2002.

The review panel then turned its attention to the choice of propellants. By 2002, the ASPL team had already experimented with a number of gases, including helium, hydrogen, deuterium and argon. However, the engine was thought to be capable of working with

---

[1] Plasmas are naturally "quasi-neutral," meaning that both negative and positive charges must exist in nearly equal amounts. To ensure this, the plasma will establish an "ambipolar" electric field to compensate for any charge imbalance that could be driven by external forces.

many others, including mixtures of gases. It was explained that, for propellants heavier than lithium, ICRH at the fundamental level would present an engineering challenge due to the requirement for higher magnetic field. Given their recent findings with helium, the VASIMR® team also discussed the potential of energy absorption at the second harmonic as an option to continue to explore. The team described other potential mechanisms suitable for energy addition in the booster stage that could be available, including Ion Bernstein Waves and parametric decay. These had not been experimentally explored. For heavy propellants, suitable for high thrust at low specific impulse, greater reliance could be placed on the contribution of the ambipolar potential to accelerate the plasma and less on the RF booster.

To the VASIMR® team in 2002, the use of a single propellant seemed to be optimal for a given I_{sp} range. More than a decade later, this approach seems to still hold. Nonetheless, the review panel was intrigued by the potential use of several different propellants and/or mixtures of propellants. The VASIMR® team explained that propellants or mixtures could perhaps be switched like gears in an automobile to optimally cover various I_{sp} ranges. This was merely a conceptual operational feature, which had not been explored extensively in the laboratory. However, the VASIMR® first stage in the ASPL had demonstrated good plasma production with hydrogen, deuterium, helium, nitrogen, argon and xenon. Experiments with ammonia were also planned for the future. There were, of course, concerns associated with the management of complex propellant mixtures and the radiation heat losses that heavier than hydrogen or deuterium propellants could bring, to say nothing of the higher magnetic field that would be required to accelerate them. Clearly much more research would be needed to understand this capability.

The performance of the magnetic nozzle was also of considerable interest to the reviewers. Many in the traditional electric propulsion community believed that the plasma would remain attached to the magnetic field and return to the rocket to produce no thrust. This rather simplistic notion was, of course, incorrect. Fortunately, one of the members of the review panel, Professor Roald Sagdeev, had produced an elegant and simple theoretical description of the plasma detachment physics. The process, however, was not experimentally verified until several years later by Dr. Christopher Olsen, a Research Scientist with the Ad Astra Rocket Company. The experimental device used for this demonstration was the 200 kW VASIMR® VX-200 prototype undergoing tests in the Company's 150 m^3 vacuum chamber.

Looking at subsystem efficiencies, the VASIMR® team quoted expected magnetic nozzle efficiencies between 90 and 95 percent for the 2002 review panel, with minimal to no field shaping, while also recognizing that for nozzle efficiencies greater than 95 percent the system would require field shaping, with an additional penalty due to the weight of additional magnetic coils. Such refinements would require careful study. On the RF power processor, the team quoted expected RF sources running at more than 85 percent. This has proven to be a remarkably accurate estimate, as the VX-200 RF generators have been shown to operate between 95 and 98 percent, with effective antenna efficiency of approximately 90 percent.

On the engineering of the system, the reviewers had expressed a concern regarding the end plate near the helicon antenna, which they considered to present a major loss channel for energy. The VASIMR® team acknowledged this concern but explained that, while early experiments had shown substantial energy loss to the end plate, this problem had been greatly reduced by proper design. Thermocouple and light emission measurements indicated that most of the power loss occurs directly under the antenna, with small amounts of energy being dissipated at the end plate. One panelist asked if the VASIMR® team had considered a half-torus shape for VASIMR®, to generate two backwards-directed exhaust jets. The answer to that question was yes. Such a configuration was considered and may be viable in an advanced design. The team pointed to a paper published in September 1985, where Dr. Chang Díaz and Dr. Yang briefly explored these ideas [3].

The reviewers then turned to system level discussions and overall technology readiness, wishing to know the current Technology Readiness Level (TRL) and the practicality of attaining TRL 6 in a 5–10-year timeframe. Interestingly enough, over a decade has indeed passed since that question was posed and under the Ad Astra Rocket Company, due mainly to the strong investment of private funds into the project, the current technology readiness of the engine is approaching TRL 6. Nevertheless, in 2002, the panel wanted the VASIMR® team to break down the VASIMR® system into major subsystems and describe them, characterizing and justifying the present TRL for each major subsystem. The panel also wanted the VASIMR® team to identify the enabling science and engineering advances required for each major subsystem to attain TRL 6 and to identify the most challenging tasks.

The VASIMR® team described a strategy of employing components, materials and techniques that offered the best combination of overall system performance and technical maturity at the time of system design. The presenters indicated that they would consider state-of-the-art technology but not at the expense of unnecessary technical risk. It was their opinion that, while technology improvements could bring additional gains, the system could already be built with relatively mature existing technology. The most challenging tasks were overall system integration and thermal control. The VASIMR® team summarized the major subsystems.

The propellant feed system was considered to be at TRL-9. The engine used gaseous propellant at ambient temperature and low pressure (~25 psi). The mass flow control device that would be used in a VASIMR® system would be similar to what is used on other existing electric propulsion systems, some of which are already flying in space. On the helicon and ICRH radio frequency (RF) generation, a TRL of 5–6 was estimated. For low power systems, less than 100 kW, all-solid-state RF components would be used. For high power systems, vacuum tube technology may be required. Even at the time, RF power technology was already very mature in ground-based applications for fusion research. Some technology development would be required for space flight and further technology advances were considered very likely.

The rocket structure and general architectural approach was considered to be at TRL-4–5. The general design approach involved an open architecture, which took advantage of the space environment for vacuum insulation and thermal control. Structural materials were expected to be metals and composites. Heat transport from interior areas to the radiator surfaces could be accomplished with integral heat transfer designs. Thermal control remains one of the most challenging areas. However, the greatest thermal loads, which appear under the helicon antenna where ionization is taking place, appeared manageable. Helicon tube materials and design are important and much engineering work would be required to address these unknowns. More than decade after these questions were posed to the VASIMR® team, indeed, much progress has been made in this important area.

Amazingly enough, the review panel was intrigued by the motivation for a space flight demonstration. The VASIMR® team indicated that the stated goals of the initial flight demonstration on the space station were to demonstrate the VASIMR® system's operation in the space environment (TRL-6) and to characterize its performance and induced environment. Another motivation was to use the VASIMR® experiment as a means to demonstrate the potential of the ISS as a generic facility for testing and flight certification of a variety of electric propulsion systems. There were important physics issues that could be unequivocally addressed, such as the plasma detachment, perhaps proving the detachment concept of super Alfvenic flow or other methods of plasma separation. Additionally, the physics associated with unavoidable plume interaction with chamber walls would be eliminated, thus producing a more relevant test. There were also space environment issues, such as those introduced by high speed bodies moving through the Earth's ionosphere. These in-situ conditions would be readily available for study in space. The team explained that high power EP could soon begin to severely tax the capabilities of existing vacuum chambers on Earth. While for simplicity, the proposed gas blanket concept would not be included in the first test, it would be a logical test to be carried out in the infinite vacuum of space. Finally, testing of EP devices on the ISS could provide an added bonus. If properly managed, the impulse imparted to the orbital complex would contribute to reducing the considerable re-boost costs of the facility.

The space station itself would be the ultimate thrust stand and a number of externally mounted plasma diagnostics could be utilized during the space demonstration. These would assist in assessing the flight performance of the engine, characterize the plasma that is produced, verify and quantify the plasma detachment process and understand the interaction of the exhaust plume with the environment and the ISS itself. Instrumentation would include Langmuir probes to measure plume density and electron temperature, retarding potential analyzers to measure ion temperature and flow rate, and plasma wave receivers that could be used to study instabilities and generation of plasma wave emissions. Proper diagnostic locations to be considered included mounting areas near the engine, moveable instruments on a dedicated boom and a deployable package that could be maneuvered by the robotic arm. Such a deployable package could be used to probe the farther reaches of the exhaust plume. Photometers and bolometers could be used to determine the plasma start-up and help assess efficiency. Existing plasma instruments already on the ISS, such as the Floating Potential Measuring Unit (FPMU) Langmuir probe suite, could be used to validate other instruments and monitor engine performance. Questions to be addressed by these instruments included: 1) Is a plasma discharge being produced? 2) What is the ionization efficiency? 3) What is the antenna loading and power delivery of the RF system? 4) What are the electron and ion temperatures? 5) What densities and flow rates are being achieved?

In further discussions, it was stated that other instrumentation should be included to measure the antenna loading from the plasma. A microwave interferometer could be used to determine plasma density and Laser Induced Fluorescence (LIF) and laser Doppler techniques could be used to sense ion temperature and flow speed as an alternative to probes directly inserted into the plasma. Some of the plasma waves that might be generated could be sensed externally with magnetic antennas and appropriate receivers. The VASIMR® team made the case for an integrated test of a complete system to be conducted in space, instead of only the helicon ionizer as one of the panel members suggested to save cost. While a compelling case for exploring the thermal management challenges with the first stage alone could be made, it was the intention of the VASIMR® team to demonstrate the space performance of the entire system and compare the flight data with ground measurements in the vacuum chamber.

The motivation for space testing was then – and continues to be today – multifaceted, involving issues of validation of physics ground measurements, as well as technology; for example, the integrated testing and thermal control of the superconducting magnets. It was stated that the ISS now provided a convenient, human-tended, research laboratory that was not available before. However, in order not to levy a high electrical load on the ISS power grid by the operation of a high power electric propulsion system, the initial ISS demonstration was proposed to be battery-fed at up to 25 kilowatts in pulses of up to 10 minutes. Longer pulses at lower power would also be envisioned. There would be enough propellant to conduct daily tests of several minutes over a period of 3-6 months. Maximum thrust was expected to be approximately 0.3 N at an I_{sp} of 9000 seconds. While the VASIMR® team was not proposing to demonstrate an operational thruster, these parameters were chosen in order to provide a compelling demonstration of the thruster's potential to provide a compensating force to a significant fraction of the ISS atmospheric drag.

Years after the review panel took place, the Ad Astra Rocket Company proposed a more comprehensive test of a 200 kW VASIMR® thruster package on the ISS. The chosen power level was consistent with the company's predicted optimal commercial niche for high power electric propulsion in cislunar space. Examining the technology landscape in the early 2000s, Ad Astra anticipated that high power solar arrays could mature by the end of the first decade of the 21st century and the VASIMR® engine would provide an ideal match for primary propulsion of high-payload, cislunar electric space tugs. The proposed test, called the Aurora Power and Propulsion Test Platform, was thoroughly reviewed by a technical panel from the ISS Program Viper team, who concluded that there were no technical impediments to conducting such a test. Nonetheless, budgetary priorities and less than lukewarm support for the test within the ISS management inhibited work on the program and delayed key milestones. A 2008 Space Act Agreement between NASA and the Ad Astra Rocket Company to carry out the ISS space test was allowed to expire quietly in 2013. This topic will be addressed further and in greater detail in Chapter 7.

It is illuminating to recall the discussion on the fundamental science questions that, at the time of the peer review, were considered still unresolved. The panel wanted to know which of these issues were endemic for electric propulsion systems in general and which were VASIMR®-specific. For those endemic to electric propulsion in general, the VASIMR® team mentioned the energy cost in the ionization of the propellant, the magnetic detachment issue, which could be important to other applied field devices, radiation losses from the plasma and stability. For the VASIMR®-specific issues, the team addressed the coupling of the RF energy to the plasma, the efficient acceleration of the ions in the RF booster and the magnetic nozzle. More than a decade later, these issues have all been resolved.

The RF expertise of some of the review panel members was quite well known and became evident during the review. The VASIMR® team was, however, thoroughly prepared for that line of inquiry. For example, the panel was under the impression that the density achieved on VASIMR® was lower than other helicon experiments had achieved and wanted to know why. The VASIMR® team explained that the helicon configuration in the VX-10 experiment was distinctly different from most other helicon experiments reported in the literature. The VX-10 created a supersonic (Mach ~ 1) plasma that flowed into a large volume where the background neutral pressure was low (~10^{-4} torr.) Furthermore, the VASIMR® team generally measured and reported properties of the plasma downstream outside of the source, where the flow velocity was high; therefore, the density was proportionally lower. Most measurements in the helicon literature are taken with neutral background pressures greater than 2 mtorr, so the plasma is relatively stagnant (Mach ~ 0.1). The VASIMR® team had taken measurements with a Mach probe, upstream of the magnetic mirror closer to the helicon source, which indicated a higher density and lower flow velocity. An RF-compensated probe, provided by collaborators from the Los Alamos National Laboratory, had measured ion saturation currents indicating a density in the source, where the neutral pressure was ~ 10 mtorr, on the order of $10^{19}/m^3$. These measurements were shown to be consistent with those reported by ORNL and other groups for helium. Both the ORNL and ASPL teams measured densities for H_2 and D_2 (on the same order as for He) that were higher than those reported by other groups. During his extended stay at the ASPL, Dr. Roderick Boswell, from the Australian National University, confirmed that the VX-10 experiment was operating consistent with other helicon experiments and noted that, "the density on the VASIMR® ionizer is high and comparable with other high density helicon systems, at least with argon."

A discussion then ensued regarding the "optimum" VASIMR® magnetic field required for helicon operation at maximum density with light gases, and how the field strength in the VASIMR® experiment was considerably higher than that found by other investigators. The VASIMR® team indicated that, for light gas operation, the results were comparable to those reported by other researchers. In their most recent magnetic configuration in VX-10, the VASIMR® team found that the maximum plasma production with the lower hybrid matching condition at the helicon antenna was the same as that reported by other groups. For argon, the VASIMR® team observed similar results to those seen by Boswell on the Basil experiment, where the maximum argon ion laser output was with the lower hybrid matching condition. The ASPL investigators, however, qualified that the data obtained thus far with argon had been sparse. The team suspected that the earlier results that they had reported with helium, where the magnetic field was much higher, were due to different boundary conditions. The collaborators from ORNL saw similar results in a similar configuration with the Mini RFTF and reported that the lower hybrid matching condition may not be a restriction for future devices.

The VASIMR® team further explained that the optimum antenna geometry and frequencies had not been chosen. The criteria for optimizing the helicon antenna was the plasma density, minimizing the ionization energy and maximizing the ion flux. They had found excellent helicon results with several antenna geometries and ORNL had explored antenna arrays. Further research would be needed to finalize an antenna geometry in conjunction with the magnetic and gas tube geometries. The team expected the helicon frequency to be near the lower hybrid frequency, but as stated, this may not be a strict requirement. The optimum frequency may be partially RF technology driven. For the helicon, the assumption at the time was an optimal frequency of 50 MHz, which was near the operating frequency in the experiments at that time and produced good magnetic field line contours into the booster section. Also, 50 MHz was easily achieved with RF technology, even in solid state devices. The team acknowledged the work of Eom, *et.al.* [4], where higher frequencies were being used.

On the booster antenna design, the VASIMR® team acknowledged that optimal geometries were still under study and were not as far along as for the helicon. Plasma loading continued to be the prime issue for efficiency. Cyclotron frequency matching required that magnetic field strength in the booster section be higher than that in the helicon section, a condition that led to a smaller plasma diameter, by field line mapping. Plasma loading increases with plasma size, but decreases with frequency, so there is a trade-off that must be made. Detailed calculations showed good loading with a booster frequency near 1.5 MHz. The experiment had been tuned for 3 MHz, so the plasma size was still small. The VASIMR® team indicated that experimental studies were in progress and remained consistent with calculations. It was stated that other experimental efforts were contemplated in the near-term, including increasing the size of the ionizer and also exploring electron cyclotron heating as a suitable plasma source and potential alternate to the helicon. These activities, however, required additional hardware modifications and were presently unfunded.

The discussion turned once again to the issue of plasma detachment. The reviewers pressed the notion that the physics of plasma detachment still presented a problem of practical concern and wondered if anyone had verified/validated the theory of detachment at the Alfvén velocity, which was studied by Breizman and Arefiev. The VASIMR® team affirmed that the process of detachment had not yet been demonstrated experimentally in a thruster configuration, but that the separation of plasma from the magnetic field is observed in nature in the solar wind, as well as in other laboratory experiments such as theta pinches, where the plasma pressure is generally high. The reviewers were aware that the VASIMR® team had proposed a number of mechanisms for breaking the constraint of adiabatic invariance and effecting detachment, including induced turbulence, mass injection near the exit and other options. They were, however, curious about how each of these techniques might impact both the mass utilization efficiency and thrust efficiency.

On this topic, the VASIMR® team explained that a number of preliminary experiments had been conducted in VX-10 with mass injection, a subject being addressed by one of the collaborators, Mr. Greg Chavers from NASA's MSFC. Preliminary results indicated that detachment of the momentum had been observed, but not in an optimized configuration. There were also other plans for ground tests of detachment theories, some of which were underway. However, there had been other proposals that included the use of existing facilities with high density plasmas in magnetic nozzles, such as the Gas Dynamic Trap in Novosibirsk, Russia, and an experiment being proposed at the MSFC facilities. Unfortunately, at the time of the peer review, the pumping capability of the experiment was inadequate for extracting any experimental evidence of plasma detachment in the existing VASIMR® set-up. Plans for upgrades were in place and components available. These, however, awaited proper funding.

The reviewers turned again to the topic of thrust measurements. They inquired about the status of the exhaust stream measurements being done with the new force sensor probe. In particular, the panel inquired about the thrust results and the measurement of the radial profile and sought consistency with other parameters. To this line of questioning, the

VASIMR® team indicated that the force resulting from the entire plasma beam impacting the force sensor had been determined. Also, a central portion of the beam had been measured, but determination of the radial profile had not been attempted. The results were still under study and the force measured could be used to "infer" absolute thrust. However, modeling of these calculations was still ongoing.

At the time of the review, the maturity of the force sensor was still low. However, the measurements provided a great deal of useful information. For example, a relative change in force between two test conditions was considered to be a direct performance metric of relative thrust. For a given configuration, the force imparted from the entire plasma beam plus impinging neutrals, with an estimated 30 percent ionization fraction, was found to be 5.5 mN (millinewtons) at one axial location, while the force 17 cm farther downstream was determined to be 6.5 mN. Combined with the flow rate of 100 sccm (standard cubic centimeters per minute), the I_{sp} was determined to be about 2,000 seconds. Additional force was observed when the RF booster was operated; the force scaled with power applied to the booster. The force target diameter was only 3.2 cm for these tests.

Once again, the panel turned to the uncertainty about the RF booster stage, stating that in some papers, reference was made to the suitability of the source stage alone as a thruster, while in others, reference was made to two stages, a helicon and an RF booster, with a plan to use ICRH. The VASIMR® team expressed that this was not so much an uncertainty as a statement of the design evolution, based on their increasing knowledge of the physics. First, the importance of the ambipolar potential in imparting some thrust to the plasma had become evident from theirs and others' experimental measurements. At the time of the review, the VASIMR® team was interested in exploring the operational range of the source stage as an effective thruster, particularly with heavier propellants, which may be unsuitable for ICRH. Secondly, the ICRH experiments at the second harmonic showed intriguing results: both substantial ion energy increase and an increased density. The mechanisms responsible for this behavior were not known, but the results pointed to the possibility of other means, beyond pure ICRH, which may be available for ion acceleration. To capture this general thinking, the VASIMR® team was now using the term "RF booster" instead of "ICRH stage".

The panel turned again to general discussions of engineering and technology and inquired about the main unresolved engineering questions of the VASIMR® concept and its implementation. Once again, they wished to know which engineering issues were endemic for electric propulsion systems in general and which were VASIMR®-specific. The VASIMR® team stated that the main engineering questions were to: 1) determine the optimum dimensions of the antenna assemblies; 2) completely characterize the internal thermal loads; 3) demonstrate effective heat transport and rejection; 4) identify the best propellant choices. The need for good thermal design and analysis is common to all electric propulsion systems and propulsion systems in general. Antenna design is unique to the VASIMR® system. This was stated to be somewhat analogous to electrode design for other systems.

One of the panel members commented on the "enormity" of the VASIMR® laboratory experiment and inquired if the technology had to operate on such a scale to prove or demonstrate the feasibility of all subsystems. He was particularly keen on scaling down the system in order to make it fit the historical standard thrust stand test used on traditional electric thrusters. The VASIMR® team explained that currently, the reasons for this condition were mainly cost and the team's emphasis on understanding the physics. At the time, the VASIMR® experiment was a derivative of the tandem magnetic mirror, first developed for fusion research. The first VASIMR® experiment at MIT in the 1980s was patterned after the Phaedrus tandem mirror experiment at the University of Wisconsin, to study the basic physics of ICRH plasma heating and transport. Most of the major components from the earlier device were still in use at the time of the review. The ASPL device was easily reconfigurable and allowed for major architectural changes at minimal cost. It provided good access for diagnostics and made allowances for the large amount of zero cost, surplus equipment being used in the project. The development of a compact VASIMR® device which could be mounted on a thrust stand had been a long standing goal. The activities in superconducting magnets and solid-state RF amplifiers ongoing at the time of the peer review were geared to that end, but funding continued to be the pacing item. The team explained that the fabrication and testing of such a device was part of a NASA Research Announcement (NRA) proposal, which, unfortunately, was not selected for funding.

The VASIMR® team explained that many factors contributed to the optimum size for the VASIMR® plasma; some were physics driven and others were technology driven. They addressed energy efficiency, fuel efficiency, power supply availability, mission requirements, RF coupling, defining the range of thrust, plasma energy and gas flow rate. Technological factors included available RF frequency, available propellant, RF power density limits for helicon and RF antennas and the thermal shielding needed between plasma and magnets. Physics factors included magnetic field requirements, plasma density required for efficient helicon operation and ICRH coupling. All of these factors had been included in the development of a 1MW point design presented to the reviewers.

On the topic of endurance, the panel inquired about full-power, endurance, and/or lifetime tests on VASIMR®, specifically for how long and at what power level. They inquired about which components had problems, which parts were likely to have lifetime problems in space, and why. To this line of questioning, the VASIMR® team acknowledged that none of the hardware on the current experimental device were flight, or flight-like, components. Therefore, no endurance or life testing had been performed using components likely to comprise a flight system. The high temperature superconducting magnet and the 1KW solid state amplifiers, which had recently been acquired by the ASPL were the first flight-like items to be tested and neither had yet undergone endurance testing. The existing device had, however, proven to be very robust, having operated at the time for several thousand hours without a failure of a critical subsystem. The principal source of unreliability had been vacuum feed-throughs, which would not be required on a flight system.

The VASIMR® team stated that, since the propellant never physically contacts engine components, the propulsive portions of the system are likely to have quite long service lives with low failure rates. Adequate life and reliability of power supply and conditioning, command and control, thermal and RF subsystems would be assured by measures such as active or stand-by redundancy, parts derating, adequate environmental protection and other well understood reliability assurance methods. This design effort would also be supported by extensive life testing of the flight system or systems.

The review panel continued to probe into failure-mode chains, such as sudden loss of propellant gas flow reducing RF coupling efficiency which might cause damage to the antennas or power supplies. To this question, the VASIMR® team stated that failure propagation was a consideration for the existing experimental device, and would be considered in a detailed flight system design. For instance, the high temperature superconducting magnet power supply was quench protected (via "crowbar") so that a magnet problem would not be propagated to the power supply and would not result in ongoing damage to the magnet. The RF system components were likewise protected from anomalies in the propellant or magnet systems. Flight systems design would include equipment to detect failures, and would provide fast, safe shutdown of affected subsystems. A detailed Failure Modes and Effects Analysis would be developed in parallel with the systems design to discover vulnerabilities such as the ones cited earlier. The current ASPL team included reliability engineering capability.

Other questions from the reviewers addressed thermal management. They wanted to know if there had been any estimates of the heat fluxes to the walls for a full-scale high power device and, given candidate materials for the walls, what limitations these heat fluxes placed on the achievable performance. The VASIMR® team expressed that heat flux to the walls for a high power device may actually be quite small. The "open architecture" (i.e., truss structure rather than solid walls) would minimize power input to the walls of the device. The open structure would allow a significant proportion of radiated power to pass harmlessly to space. In addition, it would also maintain hard vacuum in the vicinity of the thruster core, thus minimizing the number of hot neutrals in the area, and in turn, conduct heat transfer.

There was an abundance of ideas being explored to address thermal management issues. For example, optical countermeasures were available for use on the magnet enclosures and on the central tube. Highly reflective coatings could be applied to the magnet enclosures, causing very little of the remaining radiated power to affect the magnets themselves. Material options might be available to mitigate heat flux right at the central tube. Use of a highly transmissive material for the tube itself, combined with a highly reflective coating on the exterior of the tube, would have the potential to intercept heat flux before it escaped the tube. If the tube was also highly thermally conductive, then adequate heat sinking of the tube would pass this waste power directly to the thermal control system. CVD diamond was a candidate tube material, as was discussed earlier.

The review panel turned again to the propellant issue, stating that the use of hydrogen as a propellant as well as for cooling the HTS coils seemed reasonable. The panel inquired if the VASIMR® team had considered any other liquid coolants or other means of cooling the magnets, should hydrogen not be the propellant of choice, and how that change might affect the system weight. To this, the VASIMR® team stated that the use of cryogenic hydrogen had been considered for cooling of the superconducting magnets, for cases in which the engine was part of a free-flying spacecraft using that propellant. However, the design of the initial space demonstration for the ISS and for the 1-megawatt engine did not require cryogenic propellant for cooling. The design made use of closed-loop cryogenic coolers and this system was included in the overall mass estimate.

That line of questioning led to a discussion on the superconducting magnet, specifically on the power supply requirements to drive it. Given that the magnet operated on DC current, the reviewers wanted to know if the VASIMR® team had estimated the system weight, including AC to DC conversion for those power systems that generate AC electric power. The VASIMR® team stated that, in the case of the 1-megawatt engine design, they were assuming a low-power DC supply for the magnet. On a large multi-megawatt spacecraft, there would likely be a variety of spacecraft systems which would require such power. A magnet power supply item was included in the mass estimate that the VASIMR® team provided to the reviewers. In the case of the ISS demonstration design, a magnet power supply with appropriate power conversion was included in that mass estimate.

The reviewers further questioned if there were magnetic field changes required for CPT and/or maneuvering at a rate high enough to influence AC losses and thus heat load and stability of the coils, not to mention the required voltage from the power supply. To this, the VASIMR® team replied that the changes in engine performance parameters would be very gradual over the course of a mission. No serious problems with changes in magnet current settings were expected. The panel also wanted to know if any analysis had been done to determine whether all coils should use a common cryostat, or whether separate cryostats would be preferable. The VASIMR® team stated that they had considered both options. In earlier designs, the magnet coils extended over a large portion of the length of the thruster and so all of the coils were assumed to be wound on a single spool. It made sense in this case to have a single cryostat. In more recent designs, it was stated that the VASIMR® team used a series of more compact coils, suggesting it might be better to have separate cryostats for each coil. Separate cryostats might allow for better heat rejection from the engine tube, simplify engine assembly and decrease susceptibility to total system failure. This issue would be studied more deeply in future design work.

The reviewers next inquired about the radiation effects on heat load to the cryogenically cooled coils and possible damage to the superconductor and electrical insulation materials in light of expected (external) radiation environment at the magnets. To this, the VASIMR® team stated that the magnets would be shielded from thermal radiation coming from the plasma and the external environment. Radiation from the cosmic background and other sources would not be a major concern for the high-temperature superconductor material. Some recent data [5] indicated that the current-carrying performance of these materials actually improved when exposed to radiation. Thermal and electrical insulation materials would be carefully selected in the detailed design phase, for tolerance to radiation and other hazards in the space environment. The magnets would also have to be thermally shielded from the heat radiated by the external surface of the rocket tube. The magnets and the magnet cryostats would not be directly exposed to the plasma. The magnets would be located some distance from the power processing section and structurally isolated from the higher temperature portions of the engine system.

The reviewers inquired about how the VASIMR® team envisioned energizing the superconducting magnets. For long missions, heat leak can be minimized if the coils can be operated in persistent mode, but it seemed that for mission flexibility and CPT, it would be necessary to change currents during the mission. The panel was curious about the type of current leads used, such as vapor-cooled conventional or High Tc, whether they would be removable or fixed and, if removable leads were used, the issue of persistence (assuming HTS coils). They wanted to know the decay rate achievable at that point, and its projections into the future. Addressing this question, the VASIMR® team expressed that the

1-megawatt engine design and ISS demonstration design assumed a continuous power connection to the magnets. However, if possible, they would use a persistent mode with an operational system, since that would lower the cooling requirements. It would be desirable to have a re-connectable system, so that the coils could be re-energized as needed and occasional changes to magnet current could be accomplished, if necessary. These issues would be addressed in a detailed design phase.

The panel inquired if the magnets would be connected in series with the same operating current, or whether system optimization preferred different coil currents. If there were multiple currents, were the issues of total heat leak and number of components (leads) included in the weight estimations? Also, the panel inquired whether the magnet systems required quench protection and if so, how was that achieved and were those components also included in the magnet system weight estimate? To this, the VASIMR® team indicated that the present design assumed a separate power supply for each magnet. The requirements for separate leads were included in the weight estimates and quench protection would be included in the magnet power supply unit.

The review panel also received a complete list of the VASIMR® team's publications for the last 10 years. These were organized into: 1) publications in peer reviewed journals; 2) conference proceedings, including papers presented under sponsorship of the American Institute of Aeronautics and Astronautics (AIAA); 3) other reports and technical documents; 4) conference presentations.

Finally, the panel inquired about the VASIMR® team's organization, including how the team was organized to address the technical issues of component, subsystem and system integration and how the VASIMR® team would characterize the status of component, subsystem and total system design and optimization. The VASIMR® team stated that it had a flexible and operationally oriented organization, with minimal administrative structure. At the component level, technical expertise in physics from academia and national laboratories was transferred into engineering design. Physics optimization was done in the laboratory. Prototype component fabrication was out-sourced as needed. Lab testing was conducted and components were baselined. This process had been successful in both the design and fabrication of the prototype superconducting magnet and with the miniaturized 1 kW solid-state RF amplifier module. At the sub-system level, a similar approach was used. Additional expertise in subsystem design was obtained on a collaborative basis from private industry. This approach was used with Lockheed Martin Inc., (Michoud Operations) in the design of a conceptual cryogenic hydrogen storage for a VASIMR® free-flyer space demonstration. At the system integration level, the VASIMR® team utilized the services of the Integrated Design Team of the JSC Engineering Directorate to develop the preliminary integrated ISS demonstration concept. In general terms, the component, sub-systems and total system design and optimization status were in great need of development resources to move rapidly to technological maturity.

REVIEW CONCLUSIONS AND THE WAY FORWARD

The review panel concluded their line of questioning with a request for future experiments and tasks being planned by the VASIMR® team. To this request, the team described the main objectives and milestones for the following two years, which included:

demonstration of ICRH; high power operations at greater than 20 kW; characterization of plasma parameters and system optimization; demonstration of superconducting operations; systems integration of flight-like components (RF amplifier, superconducting magnet); and development of laboratory demonstrations of a viable thermal management for a flight experiment.

The experiment plans for the six months following the review (2003) included: double the helicon power to 6 kW; design and install the ICRH antenna; measure ICRH loading and conduct ICRH experiments, followed by full power helicon and ICRH experiments at greater than 10 kW. Along with these plasma experiments, the VASIMR® team planned to continue the superconducting magnet development for its eventual integration into the VX-10 experiment. The review panel concluded the meeting at the end of the day and retired into private deliberations the rest of the evening.

On November 22, 2002, the ASPL team received a summary briefing from the review panel on its findings. Two of these stood out above the others:

1. *"The physics invoked by the concept is sound"* and
2. *"The technology required, given reasonable extrapolations, supported development."*

Other findings simply pointed to the need to continue the technology maturation in order to buy down risk. However, for Dr. Chang Díaz, there was one baffling finding:

"An on-site chief scientist responsible for overall planning and direction is needed. This person should be active in both the plasma physics and in-space propulsion communities."

To his dismay, it appeared that his role on the project for more than two decades had been totally invisible to the reviewers. This issue nagged at him for years. Those who were close to him and understood his role concluded that the panel likely did not appreciate his dual role during his 25 years at NASA and viewed him strictly as an astronaut whose job was solely to train and fly in space. The panel was clearly unaware that, for 23 years, Dr. Chang Díaz had actually been working two full-time jobs for the price of one.

In the months following the peer review, it was clear that those who had sought to shut down the VASIMR® project had failed. But following on from the panel's baffling recommendation that the project needed a leader, the JSC management decided to appoint a Project Manager. They chose Ms. Elena Huffstetler, a JSC engineer with excellent engineering credentials but no prior experience in plasma physics or in-space propulsion. Also, though no stable funding source had yet been identified for the continuation of the project, the management of the laboratory was moved from the Flight Crew Operations Directorate to a newly formed JSC organization called Astromaterials Research and Exploration Science (ARES). Dr. Chang Díaz was also assigned to ARES to continue to direct the ASPL, as a "Management Astronaut."

These changes in the management of the ASPL were communicated to Dr. Chang Díaz during a tense meeting with Center Director Howell. To Dr. Chang Díaz, the new assignments were indicators of a number of troubling trends affecting his position at NASA, which he could not help but add to the rejection, by the personnel office, of his promotion to a senior executive grade. The promotion had been recommended more than a year earlier by then JSC Center Director Abbey. Dr. Chang Díaz had begun to feel an undercurrent undermining his upward mobility within the agency, which flew in the face of his four Distinguished Service Medals – the agency's highest honor – awarded over the course of his NASA career.

These not-so-subtle messages helped gel his growing desire to leave the agency altogether and pursue the VASIMR® project from the private sector. He had achieved a world record seven space missions and, given that there were many astronauts who had yet to fly at all, was unlikely to be assigned to another. The new ARES Directorate was led by Dr. Steven Hawley, also a Management Astronaut, who had previously been head of the Flight Crew Operations Directorate. Dr. Hawley was a good friend of Dr. Chang Díaz and a supporter of the VASIMR® project. Back in 1986, he and Dr. Chang Díaz had flown in space together on board the Space Shuttle *Columbia* on mission STS-61C. It was not clear, however, how long Dr. Hawley intended to stay in the agency. This uncertainty also had a bearing on the long-term security of the project.

The first meeting between Dr. Chang Díaz and Ms. Huffstetler was a cordial but difficult one for both of them. In fact, Ms. Huffstetler herself was surprised with the assignment. Fortunately, her personality was very easygoing and the two of them developed a cordial and professional relationship. Nonetheless, Dr. Chang Díaz was concerned with how the project management appeared to be morphing into the classic approach, where the scientist is relegated to a backseat role. In a highly technical project such as VASIMR®, this was a recipe for disaster. The familiar "let the managers manage and the scientists go back to the lab" culture, which was less prevalent at the National Laboratories and some of the more scientific NASA field centers, remained the norm at JSC. Even more bewildering was the fact that, in the eyes of the JSC management, Dr. Chang Díaz's scientific or astronaut credentials appeared to have automatically disqualified him from being the project manager. The unfortunate misconception of the peer review panel regarding his role in the project had triggered a profound and ill-conceived directional change that he saw as a major threat to the project.

In many ways, Ms. Huffstetler seemed to understand and appreciate Dr. Chang Díaz's predicament and, acknowledging the unnatural arrangement in which the two had been placed, began to take on a more supportive role. Their working relationship thus evolved into a productive one, looking for ways to tackle the project's real threats: adequate funding and long-term stability. At NASA's request, they developed a comprehensive three-year, $15 million project plan going forward, for Mr. John Mankins, who had been designated to follow the activity at NASA Headquarters. The plan, submitted on February 14, 2003, was never funded, however, alleging agency "budget constraints." The NASA management instead opted to continue with risk mitigation experiments, which in essence, reverted to continuing with the original plan Dr. Chang Díaz and his team had been following all along.

It was now late 2003 and the disruptive effects of the peer review were slowly waning. The project had survived one more crisis and, in doing so, had gained a greater measure of credibility. The VASIMR® project's fundamental scientific feasibility, as well as its technical capacity to support human spaceflight-class missions, had been established. As a result, the team secured enough funding to continue to make progress. As time went on, and the project management plan submitted to NASA Headquarters lost support, Ms. Huffstetler continued to help the team with her organizational and management skills, but gradually withdrew and turned her attention to other projects within her directorate. To this day, however, the VASIMR® team, now a private company, owes much gratitude to her endeavoring enthusiasm and helpful recommendations during a difficult moment in the project's history.

Another major breakthrough was achieved in 2003, one that addressed the peer review panel's highest physics priority item: to demonstrate the physics of ion acceleration in the VASIMR® second stage. This result was achieved both theoretically and experimentally and came in the form of two award winning scientific accomplishments by two graduate students who were conducting their PhD theses within the VASIMR® project. First, Alexey Arefiev, a graduate student from the University of Texas at Austin, completed his "*Theoretical Demonstration of Single-pass Ion Acceleration in the VASIMR® Engine*," in which he described the mechanism by which ions in the VASIMR® plasma would gain energy from the RF waves emitted by the second stage antenna. Alexey, a brilliant young Russian immigrant with an enthusiastic attitude and excellent aptitude for theoretical physics, had been recruited by Dr. Squire and Dr. Chang Díaz during their first visit to Novosibirsk, Russia, to attend the Open Systems Workshop held at the Budker Institute for Nuclear Physics in 1998. After reviewing Alexey's academic record, Dr. Chang Díaz agreed to the ASPL sponsoring his graduate level research at UT, Austin. This allowed Alexey to enroll in the PhD program at the university, where he worked under the direction of Dr. Boris Breizman, a VASIMR® team member, UT Research Scientist and distinguished plasma theorist, also from Russia. Dr. Breizman had spent a number of years at the Budker Institute in Novosibirsk and, familiar with Alexey's qualifications, was willing to serve as his thesis advisor. Alexey's thesis work was of outstanding quality and strengthened the theoretical foundation on which the engine's operating principles were based. The seminal work earned Dr. Arefiev the coveted Marshall Rosenbluth 2003 Best Thesis Award from the American Physical Society.

Also in that year, a second award-winning PhD thesis, by Timothy W. Glover, an ASPL employee and graduate student from Rice University, presented the first convincing measurements of the single-pass ion acceleration process by ion cyclotron waves in the ASPL VX-10 experiment. The measurement, obtained with an improved retarding potential energy analyzer designed by Tim, showed the clear shift in velocity space of the ion population with the addition of varying amounts of RF power at the ion cyclotron resonance frequency. This finding paved the way for the more compact magnetic topology of the current VASIMR® design as compared to earlier ones. These measurements were further verified in combination with other sensors installed in the experiment. A 2004 paper entitled *"Velocity Phase Space Studies of Ion Dynamics in the VASIMR® Engine"* compiled a more thorough exploration of the RF acceleration process, incorporating other complementary sensors and measurements which had become active in the experiment. The paper was presented by Dr. Edgar Bering, one of the ASPL team members, and won the American Institute of Aeronautics and Astronautics 2004 Best Paper Award in the division of plasma dynamics and lasers.

There were other achievements in both physics and technology. All the predictions indicated that the VASIMR® engine was naturally suited for high power and therefore the results would be more compelling as the power was increased. Experiments in 2003, with collaborators at the ORNL, showed this to be the case, as the power of the experiment was increased from 4.5 kW to 21.5 kW with no adverse plasma or system behavior. By late 2004, the power had reached 30 kW and the instruments were measuring ion exhaust

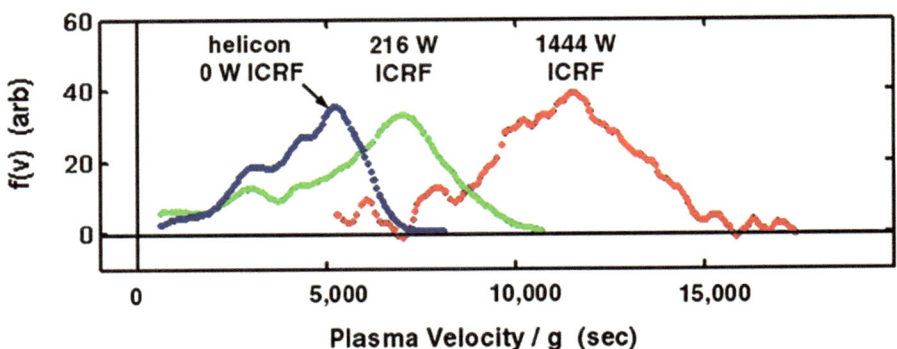

5.11 Ion flow velocities expressed in units of specific impulse for various levels of ICRF power addition.

velocities in excess of 100 km/sec on deuterium plasma. The new experiment configuration was designated the VX-50.

The advances of the VASIMR® team over eight years of research at the ASPL were clearly being noticed by the electric propulsion community at large and their results seemed to spawn a number of VASIMR®-like propulsion concepts, including the helicon double layer thruster proposed by the Australian National University, the helicon ECR thruster proposed by NASA GRC and the helicon FARAD thruster, proposed by investigators at Princeton. Work on VASIMR®-like helicon driven thrusters also began to be actively pursued in Japan and Europe.

Another important accomplishment involved the measurement of thrust. Direct thrust measurement of electric propulsion devices can be obtained by either of two methods: a thrust stand or a thrust target. In the first, the entire engine is supported on a sensitive platform or "thrust stand," which can measure a total force. In the second, a force sensor, or "paddle," is placed in the plume's path and a force is measured, which, integrated over the plume cross section, corresponds to the total thrust. While thrust stand measurements are feasible in small low power devices, they become increasingly onerous in both complexity and expense for high power systems. Moreover, thrust stand measurements require a complete thruster package in nearly finished form, which is expensive in an R&D program such as VASIMR® where physics, technology and engineering are being developed in parallel. Instead, if design remains flexible during performance characterization it can be optimized faster and with the least cost.

The much simpler and more cost effective force sensor became the preferred approach. Therefore, the VASIMR® experiment was fitted with a force sensor developed by ASPL collaborators at the MSFC. This sensor was calibrated with thrust stand measurements of a known Hall Effect Thruster in the MSFC vacuum chamber. While the helicon alone produces non-negligible thrust, its primary function remains that of a plasma source. It is the second stage booster which is primarily responsible for thrust production. The data showed the dramatic effect of the second stage in the performance of that function. In the experiment, the thrust was tripled with only an additional 15 percent burst in the input power.

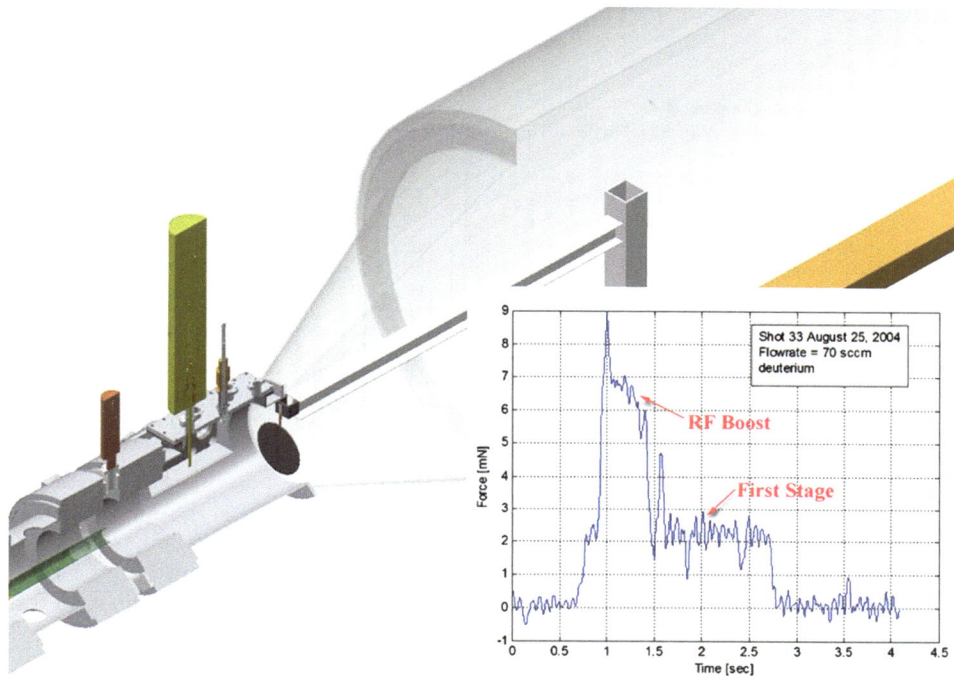

5.12 A movable calibrated force sensor was developed by collaborators at NASA MSFC and placed in the VX-10 experiment at the ASPL. The inset shows force (mN) vs. time (sec) from the calibrated sensor. With only 15 percent of the total power, the RF booster (on between 1 and 1.5 sec) triples the total thrust over the helicon first stage alone.

The use of a thrust sensor, however, assumes that the rocket plume will detach from the magnetic nozzle of the rocket itself. This was another contentious issue voiced by VASIMR® critics, which was addressed extensively during the peer review deliberations. The mechanism for plume detachment had been predicted years before by Dr. Chang Díaz and a number of researchers, but the experimental validation of that prediction in a small vacuum chamber had not been possible. The plasma detachment phenomenon is readily seen in nature as plasma ejection from solar flares and stellar jets. Nonetheless, for the VASIMR® team, a clear demonstration of plume detachment became another important research objective. A clean measurement of the detachment process, however, necessitated a much larger vacuum chamber and it would be several years before this capability was achieved by the Ad Astra Rocket Company. Once the facility enabled it, however, the detachment process was thoroughly characterized and experimentally demonstrated; this time by yet another outstanding PhD thesis from Christopher Olsen, an Ad Astra employee and graduate student at Rice University.

In late 2004, technology was another important pursuit at the ASPL. First and foremost was the development of a high temperature superconducting magnet design that would be sufficiently light, compact and powerful to meet the requirements of the plasma physics.

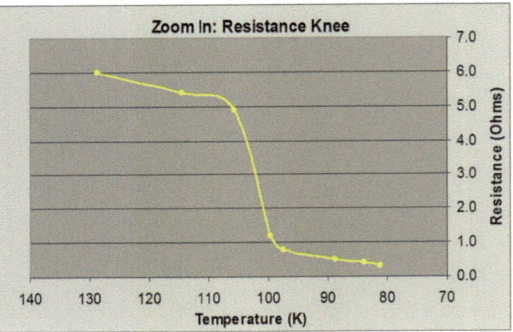

Fig. 5.13 First VASIMR® BSCCO superconducting magnet and test data showing transition to superconductivity at 105 Kelvin. The magnet operates at 40 Kelvin for adequate margin.

To do this, the ASPL team took advantage of the Small Business Innovative Research (SBIR) Program by teaming with two small American companies, Tai-Yang Research Company of Oak Ridge TN., and Creare Inc., of Hanover, NH. With that collaboration, the team successfully developed a number of superconducting magnet prototypes of increasing performance and identified the cryocooler technology needed to cool them to their operating temperature. This work continued the early research done by the ASPL team from 1999 to 2002 on Center Director Discretionary Funds (CDDF). The first Bismuth Strontium Calcium Copper Oxide (BSCCO) superconducting magnet was designed, fabricated and successfully tested at the ASPL in August of 2002.

For the ASPL, the relevance of superconducting magnet development went well beyond propulsion applications. The team initiated collaboration with the MIT group led by Professor Samuel C. C. Ting on the development of a magnetic radiation shield to protect human crews on long voyages through space. Early conceptual designs from this collaboration featured toroidal magnet structures, reminiscent of modern Tokamak fusion devices. Rather than being circular current rings, the toroidal coils would be considerably elongated so that the resulting toroid would resemble a cylindrical sleeve. Such a sleeve could be wrapped around a cylindrical habitat to provide a layer of strong magnetic insulation. The toroidal design insured that the magnetic field was totally confined within the toroidal sleeve and eliminated any magnetic field within the crew module. This feature was viewed as desirable to prevent the continuous exposure of the human crew to strong magnetic fields during their deep space voyage.

The ASPL team also explored the architecture of these magnetic structures and how they could be constructed and operated. The superconducting magnet could be nested within a cryostat where a potential engine propellant, such as liquid hydrogen or deuterium, could be stored. The liquid storage temperature of these fluids is below the superconducting temperature of advanced high temperature superconductors. Therefore, the magnetic shield could serve a dual purpose for the ship; namely, cryogenic propellant storage and radiation protection. The latter would offer the combined effect of magnetic shielding together with material shielding from the hydrogen or deuterium, which are effective radiation shielding materials in their own right. Nested within the liquid propellant tank, the superconducting magnet would not require any additional refrigeration infrastructure. The possibility of this combined approach to mitigating the radiation threat was a tantalizing consideration. However, further studies will be required to determine the technological viability of such designs.

These ideas were captured in a notional design for a 200 MW nuclear-electric Mars ship, designated as the *USS Bekuo*[2]. The ship was conceived by Dr. Chang Díaz during the ASPL years and the concept was published by NASA in the late 1990s. He introduced further refinements to the *Bekuo's* design at the Ad Astra Rocket Company, including the proper placement of radiator surfaces to support the cryo storage requirements and the nested superconducting magnetic shield. The nuclear-electric configuration would provide ample electric power to support the cryogenic propellant storage requirements. The large nuclear reactor radiators would, however, be separate structures, shown clearly in the newer *USS Bekuo* renderings. Additional modifications to the nuclear reactor radiator configuration of the ship were introduced in 2003, in order to enable high temperature radiating surfaces on both sides of the radiator structure, thus significantly reducing the radiator mass requirements.

Beyond physics and technology, another important intangible for the ASPL was that of attracting and promoting young talent and education. Research on a new interplanetary propulsion system such as VASIMR® motivates and stimulates the imagination of young people. The ASPL received a daily deluge of e-mail messages from interested prospective students and the general public. Visits to the laboratory by multiple schools and teacher groups averaged three per week, even during the summer months, and a thriving program of graduate and undergraduate student training was generating excellent results. By the end of 2004, more than 70 students from 28 different universities in the US and abroad had become involved in the VASIMR® project. The laboratory had already sponsored and shepherded seven successful PhD theses from five major universities and in that year, eight other PhD students were conducting their research on various aspects of the VASIMR®. The project hosted a similar number of Master's theses and the collaboration with teaching and research institutions generated added benefits to the project in the form of in-kind contributions of instrumentation, expertise and labor by the participants.

The growing winds of space commercialization and international competition were beginning to be felt, as more and more nations developed reliable rockets capable of reaching low Earth orbit, as well as geostationary orbital space. Fast growing China and India quickly matured their space programs and developed low cost indigenous launchers that competed very well with the more expensive US and European rockets. The Space Station MIR and the early ISS began to host billionaire tourists who would pay tens of millions of dollars for the opportunity to spend a few days in orbit. Oddly enough, the US, the ultimate capitalist nation, was slow to catch on to this new business opportunity, relying instead on the ultimate socialist nation, Russia, to show the way.

By late 2004, the ASPL had already felt the interesting possibilities offered by the private sector. Around the time of his seventh space flight in 2002, Dr. Chang Díaz engaged in some preliminary discussions with Mr. Robert Bigelow, who had come to visit the laboratory and inquired about the possibility of privatizing it under his newly formed Bigelow Aerospace Company. He had bought the rights to another highly promising NASA project,

---

[2] The name means "Shooting Star" and comes from the language of the Bri-Bri Indian tribes of southern Costa Rica, descendants of the Maya and Inca cultures of pre-Columbian Latin America.

the Transhab, an inflatable space habitat which had been developed by a team of engineers at JSC, only to be mothballed through the lack of an established agency program. The conversation with Mr. Bigelow was cordial but short, as Dr. Chang Díaz was not inclined to abandon his life-long project and was then deep in his preparations for his seventh space flight. Other interested commercial visitors came to examine the ASPL that year, including Elon Musk of SpaceX and one of the Ansari brothers, who had funded the well-known Ansari X-Prize. These visits only increased the interest Dr. Chang Díaz felt in testing the entrepreneurial waters himself and ultimately pursuing the VASIMR® engine project as a private company. There were obviously internal forces also pushing him in that direction.

It was quite evident to the ASPL team that, even for a forward-looking organization such as NASA, the VASIMR® engine was too far ahead of its time. In planning for a mission to Mars, the agency appeared to remain frozen in the 1960s mindset, promoting a national program to reach the Red Planet single handedly with massive chemical rockets. There was no priority placed on advanced technologies in space power and propulsion, which were critical to a sustainable human presence in deep space, while the agency's own electric propulsion program was in the process of being eliminated. By the end of 2004, proactively looking for a portfolio of potential solutions to the funding problem, the ASPL team submitted more than 16 separate proposals to NASA on a number of open solicitations. All of these were either rejected, or the parent program subsequently cancelled. It appeared that despite the peer review recommendations, the VASIMR® remained an interstitial object, unable to fit within the established structure. It would not be gaining a "place at the table" anytime soon and would continue to languish, as the proverbial "square peg in a round hole," in a limbo of uncertainty and instability. In private conversations with Dr. Squire, Dr. Glover and his closest colleagues and friends, Dr. Chang Díaz discussed a major strategic change to the project; a move out of the government and into the private sector. As 2004 came to a close, his mind had been made up. He would propose the privatization of the laboratory to NASA and he would leave the agency to lead the new venture.

REFERENCES

1. T. F. Yang, P. Liu, F. R. Chang Díaz, H. Lander, R. A. Childs, H. D. Becker, and S. A. Fairfax "*A Double Pendulum Precision Thrust Measurement Balance*" MIT; PFC/JA-94-26, (1994).
2. "*Single-pass Ion Cyclotron Resonance Absorption*" Physics of Plasmas Vol 8, No 3, March 2001.
3. F. Chang Díaz, W. Kruger and T. Yang "*Numerical Modeling of the Hybrid Plume Plasma Rocket*" AIAA/DBLR/JSASS Intl. Elec. Prop. Conf., AIAA 85-2058, Alexandria Va. September, 1985.
4. "*Helicon Plasma Generation at Very High Radio Frequency*" Plasma Sources Sci. Technol. 10 (2001) 417-422.
5. Roy Weinstein, private communication, 2002 and also "*The Performance of High Temperature Superconductors in Space Radiation Environments*", Michael R. James & Stuart A. Maloy, 2002.

6

A New Company is Born

In December of 2004, NASA Administrator Sean O'Keefe resigned to take a post as Chancellor of the Louisiana State University. His impending departure early in 2005 and the uncertainty about his successor were clear indicators of more management turbulence within the space agency. The changes were definitely a bad omen for the ASPL and the VASIMR® engine project at NASA and provided the final impetus for Dr. Chang Díaz to make up his mind to privatize the project and leave the agency. It was a daunting goal for a scientist astronaut and his team of researchers with no prior business experience, but it was also a seductive and tantalizing challenge, a risky move that could either lead nowhere or illuminate a bright new horizon.

Dr. Chang Díaz and his closest associates had been exploring ideas for how to accomplish such a transition for quite some time. In November of 2003, he described his concept for an "International Center of Applied Sciences for Sustainable Development" to his close friend Dr. José A. Zaglul, President of EARTH University, an international institution in Costa Rica dedicated to sustainable development. The university was in the process of receiving a substantial land gift in the country's northwest from Mrs. Marjorie Oduber, widow of Costa Rica's former President, Daniel Oduber Quirós. Dr. Zaglul was interested in exploring a mission for the new site and a multidisciplinary world-class research and training center in the applied physical sciences, critical for sustainable development, was high on the list. The concept for the center was broad in scope and could serve as a technology incubator benefiting regional development. To Dr. Chang Díaz, sustainable development went well beyond agriculture, and in his many years involved as an advisor to EARTH University he had discussed with Dr. Zaglul the importance of moving the institution beyond its traditional agricultural realm. From his experience as an astronaut, Dr. Chang Díaz equated the planet's environment with the closed life support system of a spacecraft, enabled and sustained by technology and abundant energy. In aligning space technology with sustainable development, he argued that, "if we can grow food on the deserts of Mars, we can certainly grow food on those of Earth." The proposed research center would endeavor to marry sustainability with space technology. Key areas of research included: renewable energy, hydrogen technology, bio-engineering, atomic and molecular physics, superconductivity, materials and, of course, plasma physics and space propulsion, among others.

© Springer International Publishing Switzerland 2017
F. Chang Díaz, E. Seedhouse, *To Mars and Beyond, Fast!*, Springer Praxis Books,
DOI 10.1007/978-3-319-22918-8_6

There was more than enough undeveloped land to house the project and make it succeed, but the Board of Directors of the University had become skeptical and considered the high technology mission of the new center too big a departure from the predominantly agricultural comfort zone of the institution. However, Dr. Zaglul and Dr. Chang Díaz persisted in their proposal and gradually the university board became more receptive to the idea. One important consideration was that while the university could allocate some of the gifted land for this purpose, the new venture would have to be self-sustaining and could not become a financial burden to the institution.

It was a start, but the legal mechanism to transition from the NASA laboratory to such a research facility in a foreign country was far from clear. While Dr. Chang Díaz's roots in Costa Rica were strong and the country was a long-time friend of the United States, it seemed unlikely that the US Space Agency would contemplate such a deal on ideological grounds. Even if they did, an academic-style private research facility in Costa Rica would actually be more financially uncertain than the ASPL was at NASA. More importantly, Dr. Chang Díaz firmly believed that the research had been born and nurtured in the US and was the product of an opportunity that the nation had provided when it opened its doors to him as an 18-year old Costa Rican immigrant. He had come to the United States speaking no English, with a dream, $50 dollars in his pocket, and a one-way ticket to the land of opportunity. That opportunity had materialized and had brought together the right ingredients to try out a new idea. Its resulting technology definitely belonged in the United States and, NASA's lack of interest notwithstanding, he was determined to keep VASIMR® in his adoptive country.

These and other privatization discussions among the ASPL leadership continued throughout 2004, centered around two basic choices: a private, not-for-profit research center, or a for-profit private company. After assessing the pros and cons of each, and in consultation with colleagues and friends, Dr. Chang Díaz opted for the latter. He did not see much funding viability in a not-for-profit rocket research facility outside of NASA, while seeking private investment would be far easier with a profit motive than by trying to raise sufficient capital by relying on philanthropic donors. He had also been intrigued, over a number of years, by the wave of space commercialization sweeping the US space community. The crucial question, however, remained before him: how could an active civil servant take a government facility, currently under his direction, and turn it into a private company, which he intended to lead and partially own? He clearly needed some sound legal advice.

The logical place to start was the Johnson Space Center (JSC) Office of Legal Counsel, where Dr. Chang Díaz sought the advice of Edward Fein and Daniel Remington, two NASA attorneys who had provided valuable advice to him throughout his astronaut career and had supported the legal process for securing five US Government patents bearing his name. At the same time, he also consulted with Mr. Ed Muñiz, his friend and collaborator and a seasoned entrepreneur who, years before, had walked a similar path by leaving the US Air Force to start his own engineering company.

While there appeared to be no precedent at JSC for such a transition, the reaction of the government attorneys to Dr. Chang Díaz's privatization proposal was cautiously favorable. His line of argument for a "win-win" deal for the government was simple. From the evident lack of support within NASA for a sustained development of the technology, the government seemed to be willing simply to let the project die and the research team scatter irreversibly. In stark contrast, the proposed move, at least conceptually, would be fostering the development of a technology the US government might eventually use in order to fulfill its space mission, at no cost to the US taxpayer. At the same time, spawning a new US private company was well in line with NASA's mission. Such a move could mature an important new high technology product, adding high technology jobs and commercial value to the US economy. On the other hand, if the company failed, the fate of the project would be no worse than it already was. It was indeed a "win-win" proposition, which the NASA attorneys quickly came to appreciate.

During the 2004 Christmas holiday recess, Dr. Chang Díaz studied business plans and consulted with more attorneys and business-savvy close friends about the development of a successful business. Through his friend and long-time supporter, Mr. George Abbey, now a Fellow at Rice University's Baker Institute for Public Policy, he met Mr. Robert E. Singer, a Houston attorney who took an immediate interest in the privatization concept and agreed to help mediate the proposed transition. Dr. Chang Díaz's plan was to remain as a civil servant while shepherding the ASPL through the transition to the private company and then to resign his post at NASA and leave the government service so that he could lead the new company from the private sector.

In early January of 2005, using his own personal funds, Dr. Chang Díaz hired Mr. Singer to incorporate a new company by the name of Ad Astra Technologies Inc., on his behalf. Government ethics rules prevented Dr. Chang Díaz from founding or owning a company doing business in space technology while engaged in his current line of work; therefore, in the legal transaction, there was an implicit understanding that while Mr. Singer would initially be the sole owner of the company, he would transfer total ownership to Dr. Chang Díaz once the latter left NASA.

Mr. Singer carried out his mission and Ad Astra Technologies Inc., was officially born as a Delaware corporation on January 14, 2005. The name, which is Latin for "*To the Stars,*" was suggested by Mr. Abbey and captured nicely the outward looking vision of the founders. Dr. Chang Díaz had given specific instructions to Mr. Singer to incorporate the company exactly on that date to celebrate the impending landing of the Huygens probe on the surface of Saturn's moon, Titan. The Huygens probe was a European package of instruments, ferried during a seven-year deep space voyage to the Saturn system by the Cassini spacecraft, a nuclear-powered spacecraft and orbiter slated to enter into orbit around the planet Saturn in early July of 2004. Just like Galileo a few years previously, Cassini had had to resort to gravity assists to attain the required velocity to reach Saturn. While its route was long and roundabout out of necessity, the cosmic ballet around the Sun was exquisitely timed and provided the spacecraft with two gravity boosts from the planet Venus, one from Earth and one from Jupiter. Cassini arrived at Saturn on July 1, 2004 and prepared to launch the Huygens probe for a landing on Titan, on its close approach to the mysterious moon on January 14, 2005.

The landing on Titan was a feat of extraordinary significance, probably more interesting and scientifically rich than the landing on Mars. The largest of Saturn's moons, Titan is the only one in the solar system with an atmosphere dense enough to shroud the surface features. There had been much speculation on the existence of liquid hydrocarbon lakes and rivers and the exciting possibility of intriguing processes that could be present in such a complex soup of organic matter. On that day, over one hour of discovery, Huygens descended through the thick atmosphere and landed successfully, sending to Earth stunning and never before seen images of lakes and rivers of liquid hydrocarbons from a strange world. To Dr. Chang Díaz, the successful landing on Titan was a good omen for the newborn company.

In Houston, Dr. Chang Díaz continued to develop Ad Astra's business plan and prepare a proposal to NASA for the privatization of the ASPL. As a civil servant and Director of the laboratory, his proposal would present a clear conflict of interest; however, the unusual circumstances were known and understood by the NASA attorneys, who gave permission to Dr. Chang Díaz to proceed with the proposal. It would be delivered verbally in the form of a presentation to a group of JSC officials, including his now direct supervisor, Dr. Steve Hawley, and JSC attorneys Edward Fein and Daniel Remington from the JSC Office of Legal Counsel. To stay within government ethics rules, Dr. Chang Díaz was not allowed to present his concept in an official capacity as a government employee, but rather as a private citizen on a vacation day.

The initial presentation to NASA in February of 2005 was also attended by Mr. Robert Singer on behalf of Ad Astra Technologies and kicked off four months of negotiations that ultimately led to a final privatization agreement. Although having to recuse himself from the negotiating table, Dr. Chang Díaz was, however, allowed to make inputs to the agreement draft separately as it evolved after each meeting. The process was very smooth and moved through the bureaucracy with unusual speed and surprisingly few glitches. Mr. Singer kept him updated at every step of the way as the document moved to its final stage. Alongside these negotiations, Mr. Arthur Dula, a patent attorney assisting Ad Astra in the licensing of the VASIMR® intellectual property (IP), was addressing the terms of an exclusive license agreement that would allow the new company to carry out the technology to commercialization. The negotiations for this license agreement would run well past the transition and the final agreement for the IP would not be signed until February 6, 2006.

By late spring of 2005, the final draft of the Space Act Agreement between NASA and Ad Astra Technologies was ready to sign. However, the Washington leadership vacuum existing during the negotiation process had generated some level of insecurity among the NASA management, as there had been no precedent for such a deal. A new NASA Administrator, Dr. Michael Griffin, had just taken the helm of the agency in April and the management was looking to him to stamp his seal of approval on the agreement. It was a difficult period for electric propulsion at NASA, as Administrator Griffin had begun focusing the Agency's efforts on "Constellation," his new Apollo-like flagship program to return American astronauts to the Moon. The new program had received the endorsement of the White House, but had not generated new congressional funding for NASA, and with Constellation taking center stage agency-wide, the decision had also been made that research in advanced technology projects, including electric propulsion, were to be shut down.

On June 1, 2005, a few weeks before the agreement's signature, Dr. Griffin visited JSC as NASA Administrator. As part of his tour of the center, he paid a visit to the ASPL facility, nested at JSC's Neutral Buoyancy Laboratory (NBL). Dr. Griffin was familiar with and interested in the VASIMR® project, having had a chance to learn about it years before during a meeting with Dr. Chang Díaz. Although his Constellation Program priorities would preclude the technology from being pursued at NASA during his tenure as Administrator, Dr. Griffin saw the privatization idea as a fair and novel alternative to what would otherwise amount to an irreversible shut-down. After his brief tour, his approval for the agreement was swift. Amazingly enough, in a display of ethical conservatism, the JSC management and the Office of Legal Counsel forbade the Laboratory Director, Dr. Chang Díaz, from being present during the Administrator's visit to the facility.

On June 23, 2005, Robert E. Singer, Secretary and sole legal owner of Ad Astra Technologies Inc., signed the privatization Space Act Agreement that transferred the VASIMR® project to the new company. The title of the document read:

NON-REIMBURSABLE SPACE ACT AGREEMENT

BETWEEN THE

NATIONAL AERONAUTICS AND SPACE ADMINISTRATION

GEORGE C. MARSHALL SPACE FLIGHT CENTER

AND

AD ASTRA TECHNOLOGIES, INC.

FOR

THE PRIVATIZATION OF THE VARIABLE SPECIFIC IMPULSE

MAGNETOPLASMA ROCKET

The text of the agreement defined, as a fundamental objective of Ad Astra's privatization: "…to enhance the possibility for the technology to continue to be developed to the point where it will be viable for commercialization. This will create future potential benefits for NASA as well as for those who provide the capital, time, effort, and commitment to bring the development process to fruition." The agreement also listed a number of conditions that justified the commercialization of the VASIMR® engine. These included:

1. Both NASA and related commercial development efforts for VASIMR® could benefit by developing past the current Technology Readiness Level (TRL).
2. Transitioning this technology to the private sector aligns with the Aldridge Commission's final report entitled *A Journey to Inspire, Innovate, and Discover*, presented to President George W. Bush, and recent White House space policy announcements recommend a greater participation of the private sector in space. The Aldridge Report states: "That NASA recognize and implement a far larger presence of private industry in space operations with the specific goal of allowing private industry to assume the primary role of providing services to NASA, and most immediately in accessing Low-Earth orbit." [1]
3. The VASIMR® technology has matured beyond basic research and science and is poised for rapid commercial development and deployment.
4. Emerging commercial opportunities in the private financial sector for space-related technology will attract the required developmental capital.
5. Significant and varied opportunities for commercialization of the VASIMR® technology are in the private sector.
6. NASA will use the VASIMR® as a "test case" for implementing the recommendations of the Aldridge Commission and related space policy.
7. NASA will gain from the privately developed VASIMR® technology for use in supporting its missions.
8. NASA stands to gain from the future research and development activities of the company and will maintain competency in understanding the intricacies of the new enhancements to the technology."

The agreement was co-signed a day earlier, on June 22, 2005, on NASA's behalf by Dr. Ann F. Whitaker, Director of the Science and Technology Directorate at the Marshall Space Flight Center, a reminder that while the project had been born and raised at JSC, NASA MSFC considered it to still be within their sphere of control. The signature in Houston took place, without fanfare, in a small room at the JSC Badging Office and was witnessed by Daniel Remington, NASA JSC Deputy Chief Council, who had led the negotiations that spring.

6.1 June 23, 2005, NASA JSC Deputy Chief Council Mr. Daniel Remington (left) witnesses the signature of the privatization agreement by Ad Astra's sole shareholder and director, Mr. Robert E. Singer.

The privatization agreement was very favorable to the new company and it was clear that NASA wanted the deal to succeed and for investors to be able to see the working facility and appreciate the full capabilities of the research team. To achieve this objective, it was clearly important to minimize the disruption to the laboratory across the transition. Therefore, the space agency allowed the company to continue to operate in the ASPL facility for up to 18 months after the signing of the agreement. No dismantling of the government hardware was required, the totality of which had been "loaned" to Ad Astra as part of the privatization. The Agreement was non-reimbursable, so rent and utilities during the 18-month period were also not required. All supplies, consumables, equipment maintenance and, of course, personnel costs, were now to be borne by the company.

A PAINFUL SEPARATION, A TIME TO LOOK FORWARD

The official transition of the laboratory to Ad Astra took place on July 1, 2005, producing an awkward, but brief period of team friction and uneasiness. There were four civil servants, including Dr. Chang Díaz who, under NASA's full cost accounting, had to be reassigned back to their parent organizations. It was evident that some of these civil servants – Dr. Chang Díaz included – wished the agency had fully embraced the VASIMR® project, retaining it within its fold. These individuals' enthusiasm and hard work had contributed greatly to the project's technical success. Unfortunately, despite much effort, such wishes had clearly not materialized and now they were saddened and frustrated to see the project go away from NASA and their participation discontinued. For these individuals, the security of a government job, compared to one in an as yet unfunded company, provided an easy, albeit painful choice to stay with NASA. Dr. Chang Díaz, of course, had made a different choice. During that short week, he initiated his separation process from the government and resigned from NASA on July 8, 2005, after exactly 25 years of service. The Muñiz contract with NASA was also terminated and Ed Muñiz, in an unforgettable gesture of friendship and team solidarity, agreed to retain the employees on his company's payroll while Ad Astra worked to secure enough private investment to transition the team to its own payroll account.

For Dr. Chang Díaz, his separation from the Astronaut Office brought him extreme sadness and he wanted to get it over with as quickly as possible. He was not much for teary goodbyes or flowery ceremonials. He had made up his mind, committed to a new dream, and there was no turning back. Nonetheless, he was leaving a job he had pursued all his life and had ultimately achieved, finding it to be what he always thought it would be. He had reached his dream, overcoming all the odds, in a foreign land which had welcomed him and rewarded his long struggle. In the end, his cup had "runneth over" and he was thankful to a great nation, his adoptive land, but sad that it had to come to an end. On that day, a powerful crescendo of emotions began to suddenly build up. His NASA chapter was over and, to him, a clean and sudden separation was the best course to handle the pain. Therefore, he just walked into his old office, avoiding much eye contact, and retrieved a few of his belongings, leaving most of his papers, NASA flight documents and materials behind. He accepted no farewell gathering from his colleagues the following Monday, as was customary with departing astronauts, as he felt it would be too difficult for him to maintain his composure during such an event. His NASA career had come to an end, but not his space journey. There, a new chapter was about to begin.

Things began to happen quickly afterwards. In the evening of his last day at NASA, Dr. Chang Díaz held an informal Ad Astra meeting at his home, where the company founders celebrated the successful transition. A formal meeting was held a week later, on July 15, where Dr. Chang Díaz signed the subscription certificate making him the sole shareholder of Ad Astra Technologies Inc., and Mr. Singer resigned from his interim post. After obtaining due director disclosure documentation from each candidate, Dr. Chang Díaz appointed the first Ad Astra Board of Directors which, in addition to himself, included Dr. Jared P. Squire, Dr. Timothy W. Glover, Mr. Robert E. Singer, Mr. George W. S. Abbey and Mr. Edelmiro "Ed" Muñiz. Dr. Chang Díaz was elected Chairman of the Board and Chief Executive Officer, Dr. Squire became Treasurer and Mr. Singer became Board

Secretary. On that day, Dr. Chang Díaz also injected $50,000, the sum total of his liquid family savings, as the first investment of private capital for the company. These funds were needed to purchase the insurance coverage required by the NASA agreement, as well as other company expenses.

From this point on, the world of raising private capital became Dr. Chang Díaz's new universe and operational realm. Some high net-worth individuals in the United States became interested in the new company, but the news of his resignation from NASA and his new private venture were much more widely covered in his native country of Costa Rica and he traveled there and to the rest of Latin America on multiple occasions to explore investment interest in the new company. His close friend, Dr. José Zaglul, a consummate fund-raiser in his role as President of EARTH University in Costa Rica, put him in touch with potential university donors who might also be interested in becoming Ad Astra investors.

The turning point in the capital raising effort, however, came in early August when Dr. Chang Díaz met Mr. Stephan Schmidheiny, a Swiss billionaire investor and entrepreneur with a major footprint in business and philanthropic foundations in Latin America. Stephan was deeply committed to the environment and marine conservation and also had a keen interest in advanced technology. He had become interested in the VASIMR® engine project through Dr. Chang Díaz's close friend, Mr. Carlos De Paco. Mr. De Paco and Dr. Chang Díaz were close enough to be brothers. They had grown up together in Venezuela and Costa Rica through a life-long friendship between the two families that originated as far back as the late 1930s. Carlos was a Marine Biologist who had also completed graduate studies at Harvard and was now supporting Stephan's activities in marine conservation in Costa Rica. They had become good friends and Carlos felt that Stephan would be a potential early investor in Ad Astra.

He was right. Dr. Chang Díaz met Stephan in San José, Costa Rica, and traveled with him to his large eco-sustainable farming complex, La Pacífica, near the City of Cañas in the country's northwestern province of Guanacaste. The two men had time to get to know each other, as Mr. Schmidheiny showed Dr. Chang Díaz their ongoing tilapia farming and reforestation projects. They spoke a great deal about energy, one of Stephan's favorite subjects, and gradually got deeper into the field of space and its evolving commercialization opportunities. On one of the evenings at La Pacífica, Dr. Chang Díaz presented Stephan with a comprehensive talk on his rocket development program, the privatization agreement from NASA and the elements of a long-range business plan for commercial in-space transportation, beginning with robotic cargo vehicles running on solar-electric power and VASIMR® propulsion and serving a growing worldwide space logistics business. Looking further into the future, he described the evolution of the technology all the way to multi-megawatt nuclear-electric power, enabling the fast and economically sustainable transportation of humans into deep space.

The following night, after dinner, Stephan took Dr. Chang Díaz aside and indicated his interest in participating in the project and agreed to a starting investment of $6 million, a precious and immediate boost to the fledgling start-up company. The deal was consummated in no more than a few minutes of clear and simple terminology and common vision and sealed with no more than a firm handshake; no complicated negotiations, no lawyers, no fuss. Dr. Chang Díaz was impressed with Stephan's clarity of thought and immensely proud and grateful for the trust he had placed in his new dream. He was also grateful to his life-long friend, Carlos De Paco, who had laid the groundwork and opened the door to this opportunity. Shortly after the meetings at La Pacífica, Stephan and his son Alex, along with Carlos De Paco and others in Stephan's staff, accompanied Dr. Chang Díaz back to Houston where they could view the VASIMR® engine hardware and tour the laboratory at the NASA Sonny Carter Facility.

6.2 On August 10, 2005, Mr. Stephan Schmidheiny (center) visited Ad Astra Rocket Company at the NASA Johnson Space Center in Houston. He was accompanied by his son Alex (far right) and Mr. Carlos de Paco, Dr. Chang Díaz's close friend.

The Schmidheiny investment in Ad Astra represented an injection of capital on a scale never before seen during the project's 25 years at NASA and clearly validated the premise on which the transition from NASA was based. The privatization concept was working. The money began to flow in six installments of $1 million starting in October of 2005, just over three months from the NASA transition. To the VASIMR® project, it felt like "a shot to the vein." Shortly after Stephan's investment, other large investors came to the table in the US. In Costa Rica, the news of the new US company, led by their national hero and with its bold plans to also establish an Ad Astra subsidiary in Costa Rica, ignited the enthusiasm of the nation and soon a group of high net-worth Costa Rican families joined the investor mix.

The young company was definitely on a roll. The availability of significant cash reserves in Ad Astra's bank accounts enabled the team to upgrade old equipment, do long-range planning and prepare to move fast on the remaining physics unknowns, as well as increasing the technology readiness level (TRL) of the rocket. In late August 2005, the company changed its name from Ad Astra Technologies Inc., to Ad Astra Rocket Company, a new name which was more appropriate to the mission the team had set out to accomplish. The new name also required a new logo, which Dr. Chang Díaz designed en route to Spain on October 22, 2005. The new design, which remains in use today, features the half crescent silhouette of a ringed planet making the letter D. The edge view of the ring appears as a line leading to a star, Ad Astra's ultimate destination.

Experiments continued in earnest at the new company. The Oak Ridge National Laboratory (ORNL) team, with the support of the Fusion Energy Division led by Dr. Milora, had continued to work on the physics of plasma acceleration by ion cyclotron waves in the VASIMR® engine and had proposed to loan Ad Astra an FRT-85 radio frequency (RF) generator, capable of up to 200 kW of RF power in the FM band. The system could be incorporated into the existing VX-50 configuration with a more robust ion cyclotron resonance heating (ICRH) antenna. It was anticipated that the dense plasma already possible in the VX-50 would absorb the ion cyclotron waves efficiently and produce significant plasma acceleration.

The experiment was successful. The injection of up to 20 kW of ICRH power in initial experiments with deuterium plasma produced a clear plasma acceleration signature, with bulk plasma flow moving at over 100 km/s. The VX-50 further demonstrated that the ICRH process could also be applicable to heavier gases, such as argon and neon, to generate higher thrust at a lower specific impulse. The ability to vary the thrust and specific impulse in this manner was one of the important operational features of the VASIMR® engine as compared with other electric propulsion concepts.

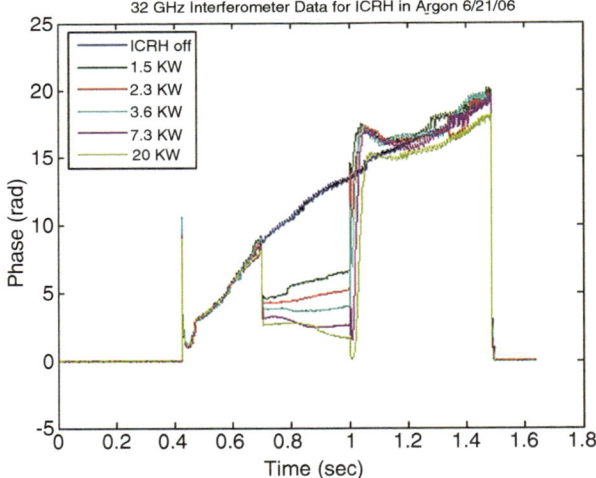

6.3 In the fall of 2006, up to 20 kW of RF power in the form of ion cyclotron waves was injected into a high density argon plasma in the VASIMR® VX-50 experiment. The RF power produced a clear indication of plasma acceleration and demonstrated the operation of the rocket with heavier gases.

There was also another important achievement in system performance; a reduction of the energy cost of ionizing the propellant. As described earlier, the ionization cost is an important parameter affecting the engine efficiency and it must be as low as possible. This value had continued to come down with the higher power helicon experiments, from 500 eV/ion in November of 2004 to 200 eV/ion in January of 2005, a number which was entering the acceptable range of efficient VASIMR® operation. While these results were exciting enough, they were conducted with deuterium, a light gas and a member of the hydrogen family of isotopes, which is notoriously difficult to ionize. As the team switched to heavier gases, such as argon and neon, the real payoff soon became evident, as the reduction in ionization cost to levels as low as 100 eV/ion was demonstrated.

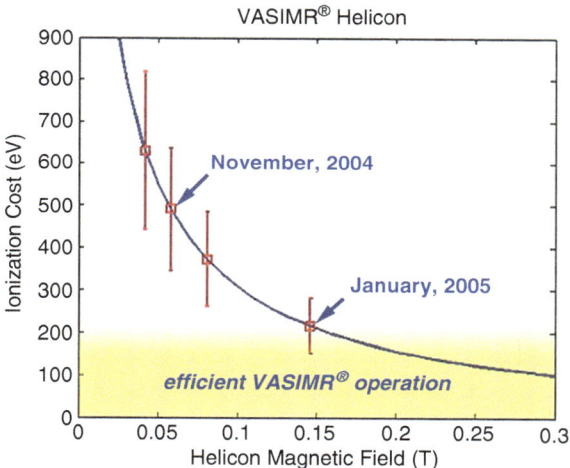

6.4 Ionization cost in eV/ion for deuterium.

SOLE SURVIVOR

On Monday October 31, 2005, several hundred rocket scientists from all over the world converged at Princeton University to attend the 29th International Electric Propulsion Conference, a biannual gathering of the electric propulsion community to present their latest advances in the field. That year, however, it was a gloomy crowd that gathered at Princeton on Halloween day, as news of the cancellation of all major US programs in electric propulsion was announced in an opening ceremony that became known as "Black Monday." There was a curious exception, however, as the newly formed Ad Astra Rocket Company contingent, led by Dr. Jared P. Squire, presented their latest results and plans to an astonished audience. The VASIMR® engine's high power helicon had achieved excellent results and the project was now moving forward under private funding to high power ICRH demonstration the following year. On "Black Monday," the VASIMR® engine, the known orphan of electric propulsion, was the only project that had survived.

There was one more important action remaining before Dr. Chang Díaz could consider the company complete: it had to support its own payroll. The ASPL team was still on the Muñiz Engineering payroll, but Ad Astra had refunded MEI for personnel expenses incurred during the transition and was now funding the MEI team as company contractors. However, the plan was to formally transition the group to Ad Astra's payroll by the following spring. The transition occurred in February of 2006 without any complications. The relationship with MEI had been a long and productive one and would continue to remain so. It was no accident that its President and founder, Mr. Ed Muñiz was a member of Ad Astra's Board of Directors.

Another key individual joined the Ad Astra ranks from the ORNL in March of that year. His name was Dr. Mark D. Carter, a brilliant scientist and engineer. Dr. Carter had been deeply involved in the VASIMR® project during the NASA days as part of the ORNL group who collaborated with the ASPL. Initially trained as a nuclear engineer, Dr. Carter had moved to plasma physics and fusion during his PhD research at the University of Wisconsin under the guidance of world renowned plasma physicist Noah Hershkowitz. When he joined the VASIMR® Oak Ridge collaborators, he was part of ORNL's theory group and had become an expert in the physics of RF wave absorption in magnetized plasmas. This was a key area of investigation for the VASIMR® team. Dr. Carter had predicted the efficient absorption of RF power in the VASIMR® engine and was eager to see it experimentally demonstrated. He had also become enthusiastic about the new private venture and expressed his desire to resign from his post at ORNL and relocate his family to Texas to join the Ad Astra team. The company now had the resources to make him an offer, which he readily accepted.

By the spring of 2006, the infusion of new capital had done wonders to Ad Astra and to the project. Morale was high and the sheer scientific muscle and experience of the team was unquestionable and was reflected in the prolific output of excellent experimental data. NASA had continued to do its part to help the company succeed. Operating a private company within a federal facility had its challenges, including the issuance of proper badges to visitors that had to cross NASA installations to get to Ad Astra. There was also the compliance with company rules that could be slightly different from those of NASA. There was a tendency for the facility operators of the building to forget that Ad Astra was not a NASA installation any longer and therefore entry by NASA personnel to the company's facilities had to be approved by the company. Nevertheless, the symbiotic relationship worked well for as long as it needed to. It had virtually eliminated disruption to the project during the privatization transition and had enabled the team to go right to work on the technology. It also had the added value of having real hardware operating in a real laboratory, as the company launched its first capital raising marketing campaign. The NASA genesis and present housing gave rightful legitimacy to the project in the eyes of potential investors, and the laboratory was something to behold, especially during active plasma operations. Indeed, many investors had told Dr. Chang Díaz that the facility tour and the view of the bright and powerful plasma exhaust through the vacuum chamber viewport were strong marketing elements for Ad Astra.

There was more still to come. For Dr. Chang Díaz, the establishment of an advanced research center in his native country of Costa Rica had been a dream for many years. He believed that space technology offered an extraordinary potential for sustainable economic growth in the developing world. The global space operations market in the early 21st Century already topped $300 billion and was surely poised to grow, as more space-borne services such as GPS and remote sensing satellites were deployed and commercial ventures began supporting the growing human presence in orbit. Moreover, space technology did not require strategic natural resources or a large domain of sovereign territory. Only brains, money and the willingness to use them were needed to enter the club of spacefaring nations.

Costa Rica was a peaceful and stable country that had made some important investments in education in the late 20th Century. These investments were already returning dividends in the form of a young and well-educated workforce that had allowed Costa Rica's economy to grow and diversify from its primarily agricultural base and to make some initial strides into high technology. At the turn of the 21st Century, Costa Rica had expanded its economic matrix into the manufacture of hundreds of new advanced technology products in the medical and microelectronics fields, alongside an extremely well managed eco-tourism industry that had grown from the nation's strong emphasis on environmental protection. Despite a lingering antiquated legal system and poor governmental management, the country's private sector was vibrant and ready to tackle new challenges. Dr. Chang Díaz, in consultation with his close friend Dr. Zaglul, believed that the conditions were right to take the step. As soon as resources were available, he had decided to plant a small space technology seed in Costa Rica by establishing a wholly owned subsidiary of Ad Astra in his native land.

Plans to build the Costa Rica subsidiary began to gel almost immediately upon Ad Astra's creation, but the implementation of the idea would have to wait until sufficient funds could be raised. The Board of Directors of EARTH University had agreed to support the establishment of a research facility in its new northwestern land grant near the city of Liberia. However, the approval of the board did not imply a land donation to Ad Astra, or any special arrangement, but simply the right to purchase a one-hectare lot to build the new laboratory on a section of the property which the university had set aside to sell. The shared vision of Dr. Zaglul and Dr. Chang Díaz was that the presence of Ad Astra would catalyze other high technology firms to cluster around it, forming a science and technology complex in a symbiotic relationship with the university.

The company now had the money, but Dr. Chang Díaz wanted to be careful with how he chose to use the resources. His concept was a simple, medium size modern facility, built to suit and equipped with the latest advances in broadband Internet communication and electrical power. Many people questioned the choice of location, however. In the early 21st Century, Costa Rica remained a highly centralized country, with very poor roads and infrastructure. The province of Guanacaste, the second poorest in the country, was considered too remote from the capital of San José and the highlands of the central valley, where most of the business transactions were conducted. Schools were poor and adequate housing was scarce to non-existent, except for lavish summer mansions and overpriced luxury condos catering to wealthy Costa Ricans and foreign vacationers. Medical facilities were few, understaffed and overcrowded. With some of the most beautiful beaches and forests of the country, it was a paradise for visitors, but a rough place for the locals; the proverbial "I love to visit but I wouldn't want to live there" location.

Dr. Chang Díaz, however, was in love with Guanacaste, the place of many of his outdoors adventures as a youngster, hiking and hunting with his father. He saw the northwestern province as the future technological Mecca of the country, a Costa Rican "Silicon Valley," and he saw Ad Astra as a catalyst to making that future vision a reality. A strategically located international airport, catering to the growing ecotourism business, already provided daily direct flights to Houston and all major cities in the United States and Europe. The airport was just a stone's throw from the EARTH campus and the site of the proposed Ad Astra facility and a brand new terminal was in the planning stage. All in all, compared with the crowded and congested urban setting of the central valley, the place was new, fresh and simply beautiful and Dr. Chang Díaz had made up his mind that Ad Astra would locate there.

To accomplish this goal, he approached his brother Ronald, a Costa Rican civil engineer and builder who, together with his younger brother Norman, owned and operated Chang Díaz y Asociados, a medium sized construction company in Costa Rica. His brother immediately agreed to take on the challenge and began to draw up plans to meet Ad Astra's requirements and deliver a finished building by mid-2006. The company also began recruiting a small workforce of young engineers, led by Mr. Jorge Oguilve and Juan Ignacio Del Valle, both mechanical engineers educated at the University of Costa Rica and with Master's degrees from the Costa Rica Institute of Technology and Delph University in the Netherlands respectively. Years before, Mr. Oguilve had been one of the Costa Rican student interns who had completed a year-long internship at the ASPL as part of the scientific collaboration the ASPL had opened with international institutions, and his previous training at ASPL had prepared him well for his future assignment.

The Costa Rica subsidiary, Ad Astra Rocket Company Costa Rica, SRL (Sociedad de Responsabilidad Limitada), was formally incorporated on October 28, 2005. The facility was completed in record time. Its inauguration took place on July 16, 2006, before a crowd of more than 1000 dignitaries, diplomats, VIPs, special guests and the general public. The ribbon cutting was especially significant: Prof. Samuel C. C. Ting and President Oscar Arias Sánchez, Nobel Laureates in Physics and Peace respectively, were joined by Prof. Roald Sagdeev, a former reviewer and now a strong supporter of the new venture, and Dr. Chang Díaz's mother, María Eugenia Díaz, the first promoter of his dream. Dr. Chang Díaz's brother Ronald, in addition to building the facility, assumed the management of the Ad Astra Costa Rica operation on a part-time basis. Every week for nearly five years, he commuted the four-hour drive between his company in San José and the Ad Astra facility in Guanacaste. During his tenure with Ad Astra, his relentless energy and enthusiasm did wonders to build the company's image among the country's public and private sectors and as a beacon for education and motivation for the nation's youth.

The building for the Costa Rica subsidiary, Ad Astra Rocket Company Costa Rica, SRL, was built with Ad Astra resources and then sold to a Costa Rica Real Estate Fund, which then leased the facility back to Ad Astra. One of Ad Astra's board members, Mr. Ed Muñiz had advised Dr. Chang Díaz to avoid Ad Astra entering into real estate ownership at this early stage of the company, as this would bring an unnecessary financial complication and distract from the young company's main line of business. The lease option was

6.5 Four special individuals cut the inauguration ribbon: (L to R) Ms. María E. Díaz de Chang, Dr. Chang Díaz's mother; Prof. Samuel C. C. Ting, MIT Professor and 1976 Physics Nobel laureate; Dr. Oscar Arias Sánchez, President of Costa Rica and 1987 Nobel Peace Laureate; and Prof. Roald Sagdeev, U. Maryland, Distinguished Professor of Plasma Physics and former Director of the Space Science Institute of the former Soviet Union.

made possible by the real estate division of Grupo Aldesa, an investment and financial management organization with whom Dr. Chang Díaz had begun to engage in Ad Astra's business transactions in Costa Rica. The company was founded by Mr. Oscar Chaves Esquivel, a well-known and respected Costa Rican economist, who had been one of the founders of the Costa Rican Stock Exchange. His sons, Oscar and Javier Chaves, visionaries in their own right, now managed the firm and had become keenly interested in Ad Astra's business plan and its future in both Costa Rica and the rest of Latin America. As the Ad Astra facility was completed, the Aldesa Real Estate Fund purchased the facility and signed an agreement with Ad Astra for the lease of the property.

The Costa Rica facility went into operation immediately. Its main task was to establish a plasma laboratory to assist the US operation with the materials and heat transfer issues associated with long exposure of some components of the engine to plasma bombardment. To embark on this research, Ad Astra Rocket Company had to apply for and obtain the required export control approvals from the US State Department. A fairly simple plasma generator, based on the helicon first stage of the VX-50, was developed locally from basic principles. The CR facility achieved first plasma on the evening of December 13, 2006.

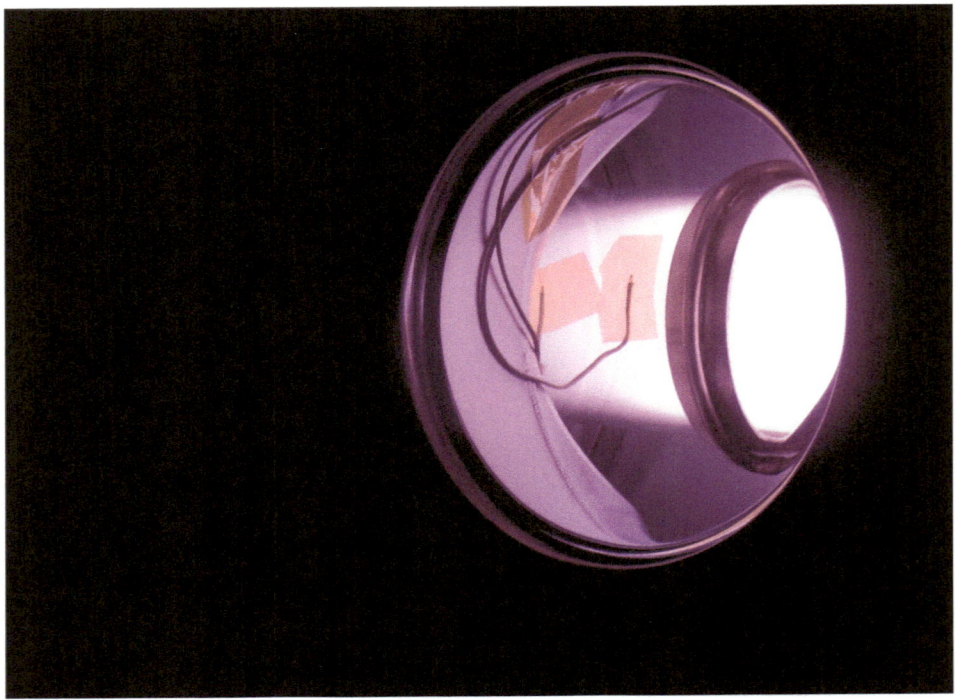

6.6 First plasma in Ad Astra's Costa Rica facility was achieved on December 13, 2006

A NEW HOME

In Houston, the old NASA VX-50 experimental apparatus was formally decommissioned on October 18, 2006 to make room for the first company-owned test article, the VX-100. This new system utilized a new water-cooled magnet, built by Everson Tesla of Nazareth, PA, to Ad Astra's specifications. The VX-100 was a transition engine, designed to test the performance of a flight-like, high power rocket core with a low cost conventional magnet. The results of this test article would provide experience and confidence for the Ad Astra team to finalize the design of a more flight-like fully superconducting engine prototype. VX-100 became operational in May of 2007 and by the fall of that year, the performance data coming from the device was exceeding expectations.

The VX-100 experiments confirmed the now widely accepted notion that the VASIMR® engine is naturally suited for high power. The test results showed that doubling the power from 50 to 100 kW actually tripled the plasma production. With these results in hand, the engineering team began developing the specifications for the next evolution of the VASIMR® engine, the VX-200, which would be fully tested in a vacuum environment and would feature the first VASIMR® superconducting magnet.

6.7 VX-100, a high power engineering test bed, was the first company-owned VASIMR®
prototype. It became operational in May of 2007.

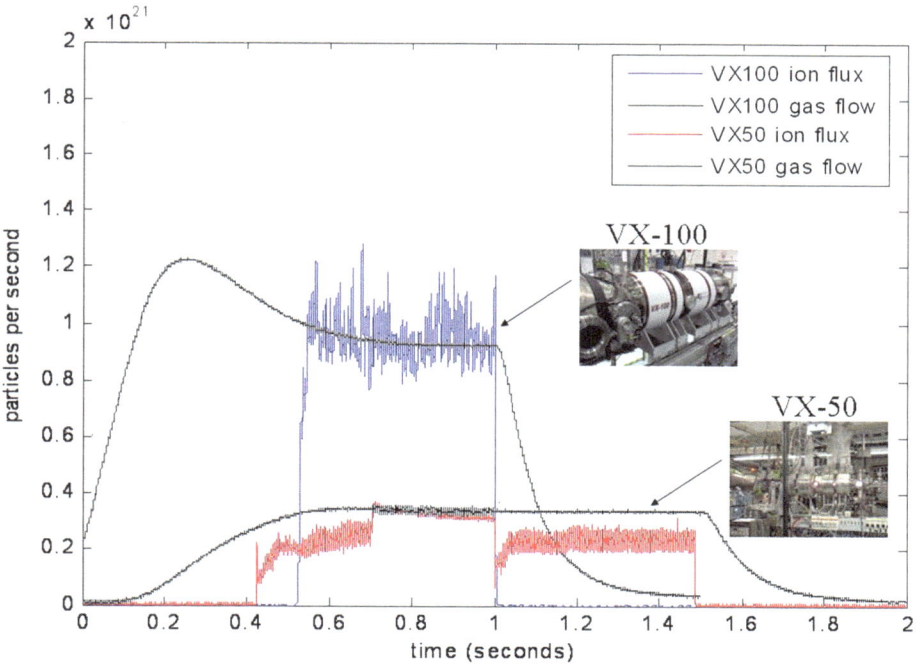

6.8 The results from VX-100 showed that doubling the power from VX-50 to VX-100 tripled
the plasma production.

During 2007, the Ad Astra Rocket Company began looking for a new and more spacious home. Up until then, the company had continued to occupy its old JSC facility at the former ASPL; however, a contract for a brand new vacuum chamber had been signed with PHPK of Columbus, Ohio. The chamber, which was much too large to fit in the NASA facility, had been designed and was being manufactured to Ad Astra's specifications for delivery in the fall of 2007. The Ad Astra leadership had considered the possibility of building a new laboratory in the Webster or Clear Lake area close to JSC; however, they ultimately opted for renting space, once again reasoning that ownership of real estate would unnecessarily encumber the company financially and introduce a distraction to the main line of business.

Many months of searching for a new Ad Astra location had produced some potential sites, but none were exactly what was needed. Other than NASA JSC, the surrounding aerospace engineering infrastructure was more focused towards providing office space for personnel, rather than physics laboratories or research and manufacturing facilities with adequate power feeds. After all, the JSC was not a test center for rocket hardware, but rather an operations and training center. High-bay building space was scarce and generally not well equipped. The housing of a large vacuum chamber would certainly introduce major demands on the high-bay, as well as the load bearing of the floor of any building Ad Astra would hope to occupy.

The engineers, however, identified an old, 24,000 square feet warehouse, located at 141 W. Bay Area Blvd in Webster, TX, about 3 miles from the JSC and hidden behind a small strip of general retail shops and restaurants. The building was old and in considerable disrepair and had no "curb appeal," but the base lease price was attractive and Ad Astra agreed to make the necessary improvements to the building at its own cost in exchange for a favorable lease agreement. The deal was accepted and throughout most of 2007, the Ad Astra leadership maneuvered through a gradual transition of operations from the old NASA site to the new Webster location. There were major adjustments that needed to be made to the new location, including the installation of up to 1.5 MW of electric power and the reinforcement of the floor to support the 40-ton, 150 cubic meter stainless steel vacuum chamber.

On October 1, 2007, the VX-100 was fired at NASA for one last time in a small ceremony, bringing more than a decade of experiments at the NASA facility to an end. The event included the ceremonial firing of the engine by several individuals who had contributed to keeping the research alive throughout all those years. One of these was astronaut Capt. John W. Young, Dr. Chang Díaz's long time hero, friend and tireless supporter, who had the honor of firing the last shot in the NASA laboratory.

With the impending delivery of the new vacuum chamber, time was of the essence and by late 2007, the relocation of the laboratory was nearly complete. The arrival of the vacuum chamber, one of the largest in the industry, was a major event for the town of Webster. The imposing, 40-ton, four-meter diameter, ten-meter long cylinder arrived in Houston on a wide flat-bed truck that had skillfully navigated the highways across several states from Columbus, Ohio, to Houston, Texas, with a police escort. On the final leg of the trip, south of Houston on Interstate 45, the truck convoy blocked most of the southbound traffic, while the final highway exit at Bay Area Boulevard and arrival at the Ad Astra facility required Dr. Chang Díaz himself, accompanied by his wife Peggy, to guide the truck to deliver its precious cargo to its future home. Many of the Webster citizens along the convoy's path stopped to witness the odd, impromptu parade of a gleaming cylindrical steel giant slowly moving through the streets, being led by a man and his girl in their silver Corvette convertible. Deputy Webster Mayor, Carlos Villagómez, had summed it all up when Dr. Chang Díaz informed him of his intention to move the plasma rocket facility to his town: "God help us!"

6.9 At the controls: Capt. John W. Young fires the last shot of Ad Astra's VASIMR® VX-100 prototype operating at the Johnson Space Center on October 1, 2007. Observing (L to R) Dr. Chang Díaz, Mr. George W. S. Abbey, Arthur Dula, Ad Astra's Dr. Benjamin Longmier, Jacob Chancery and NASA JSC Director of Technology Transfer, Ms. Michele A. Brekke.

All the modifications were completed in time to support the Ad Astra move to its new home. The inauguration of the new Ad Astra Webster facility came at the same time as the signature of a new collaboration agreement with NASA's JSC, entitled:

UMBRELLA SPACE ACT AGREEMENT

BETWEEN

NASA LYNDON B. JOHNSON SPACE CENTER

AND

AD ASTRA ROCKET COMPANY

FOR

COLLABORATIVE ACTIVITIES WITH AD ASTRA ROCKET COMPANY

This was the second Space Act Agreement (SAA) the company signed with NASA. Strategically developed by JSC's Director of Technology Transfer, Ms. Michele A. Brekke, its purpose was broadly defined to maintain a collaborative framework that would facilitate future joint activities of mutual interest between Ad Astra and JSC. Dr. Chang Díaz and JSC Center Director Michael Coats, also a former fellow astronaut, signed the document on Monday, December 10, 2007, in a small ceremony attended by a number of Ad Astra investors, members of the US Congress and NASA officials.

6.10 The second Ad Astra Space Act Agreement with NASA was signed by Dr. Chang Díaz and JSC Center Director Michael Coats on December 10, 2007 and established a broad cooperative framework for the parties to work together.

As Ad Astra upgraded its research facility with more modern equipment, it returned the loaned NASA hardware back to JSC. Included in these returned items was the VX-50 experiment apparatus. Some of its components had arrived in Houston in 1993 from Boston, including the high field cryogenic magnets and much of the early hardware that comprised the first experimental test bed of the VASIMR® engine, where many of the early physics breakthroughs had been made. The team felt that the apparatus had some historical significance and had suggested the preservation of the configuration to NASA for potential exhibition later on in a museum. Unfortunately, the property department at NASA did not appear to share this sentiment, and the apparatus was dismantled and scrapped. Fortunately, some of the innards of the early rocket that had been considered by NASA to be of no significant value to the government were allowed to stay with Ad Astra. This hardware attracted the interest of Mr. Mark Armstrong, a member of the Board of Directors of the Museum of Flight in Seattle. Mark was one of the surviving sons of astronaut Neil Armstrong, the first human to step on the Moon. He and Dr. Chang Díaz had met at an astronaut gathering in Houston to honor the memory of his father. Mark had become familiar with the VASIMR® and had inquired with his board about the possibility of including a display of the early rocket experiment in the new "Spaceflight Academy" exhibit being built at the museum. The idea resonated with the Board and for several months, its curators and exhibit developers, Geoff Nunn and Jeff Margolis, worked with the Ad Astra team to reconstruct the early VASIMR® VX-50 plasma tube from the surviving parts. The initiative was successful and in November of 2014, the Museum inaugurated its new wing which includes the only public exhibit of the early VASIMR® engine.

6.11 Dr. Chang Díaz poses by the early VX-50 VASIMR® engine hardware, on display at the Museum of Flight in Seattle.

THE VX-200

Two important new projects began to take shape in 2007. First, the team began the design and construction of the next test article, the VX-200. This system was to integrate all the major subsystems of the VASIMR® engine into one flight-like module that would also be vacuum compatible. Only the RF generators, which were being developed by Canada's Nautel Ltd, would remain outside the vacuum enclosure. This would ensure rapid access to the units during experiment operations. The solid state RF generators were rather new and the team felt it was an unnecessary expense and complication to make them vacuum ready just yet before gathering sufficient operational experience.

The second major subsystem was the cryogen-free superconducting magnet. This was a critical system and the company contracted Scientific Magnetics, a small superconducting magnet developer in the United Kingdom, to carry out the development to Ad Astra's specifications. The Ad Astra team was familiar with Scientific Magnetics from their pioneering work in the development of the superconducting magnet for the Alpha Magnetic Spectrometer (AMS), the space-borne particle physics detector that Dr. Chang Díaz had a chance to operate during one of his Shuttle flights, STS-91. While Ad Astra had chosen an advanced, lightweight, high temperature superconductor for an eventual flight engine, the VX-200 ground prototype would be equipped with a lower cost and more technologically mature low temperature unit. The low temperature unit would be heavier and would levy a higher refrigeration demand, but it would provide an excellent test case for the operation of a much colder superconductor in close proximity with the plasma in the rocket core.

There were rapid advances taking place in the field of high temperature superconductivity and the technology of reliable conductors was becoming more available commercially. Two companies in the United States led the field; American Superconductors of Devens, Massachusetts, and SuperPower Inc., of Schenectady, New York. Of the two, SuperPower seemed to have the more advanced commercial technology for the VASIMR® engine application. Their REBCO (for Rare Earth Barium Copper Oxide) superconductor, a thin, 4 mm tape, was capable of carrying about 100 amperes of electricity with no resistance at temperatures of less than 100 Kelvin. The material was also under further

6.12 Ad Astra's VX-200 being readied for vacuum fit checks in the company's 150 cubic meter vacuum chamber.

development near Ad Astra, at the University of Houston where Dr. Venkat Selvamanickam was pioneering advances in the process of manufacturing these conductors. The VASIMR® engine's requirements were not extremely difficult, given the advances of the technology, but the cost remained high compared with the more traditional low temperature wires.

The VX-200 began operations in mid-2008, with first plasma achieved on May 12 of that year, initially with an interim low field water-cooled magnet designed and built by Ad Astra. The interim experiments enabled the team to gather operational experience with the new system and the more complex protocols required in the new vacuum chamber. The superconducting magnet arrived in Houston in mid-2009, after experiencing a number of unexpected delays. The magnet employed a conventional niobium-titanium superconductor, able to reach zero resistance with sufficient current capacity below approximately 6 K. There was, however, a unique feature in this superconducting magnet, namely its cryogen-free operation. Most conventional superconductors work in a bath of liquid or superfluid helium; however, it was important for the VASIMR® engine's eventual design to build a magnet that required no fluids. The Scientific Magnetics team rose to that challenge. The magnet was carefully suspended inside its own cryostat and insulated in such a way as to virtually eliminate all heat sources to the superconducting windings. The cooling was achieved by the action of cryocooler heads attached to a thermal bus that transported all the heat away from the windings.

The magnet, of course, required a vacuum environment to operate, so the testing of the VX-200 engine required lengthy preparations. First, the entire assembly was rolled to its test position inside the vacuum chamber and pumping protocols were initiated. It would take several hours for the chamber to reach acceptable vacuum conditions, especially after lengthy periods in an open configuration where the high humidity of the Houston summer would tend to trap moisture in the various crevices of the device. Once acceptable pressures were reached, the chilling of the superconductor would begin. It would take up to three days for the cryocoolers to bring the magnet to superconducting conditions. All in all, however, once an experiment campaign began, the team proceeded with great speed. The operation of the engine and all the diagnostics were controlled from an array of computers arranged in an experiment control and command center in front of the chamber. Test procedures and objectives were well rehearsed and prepared ahead of time. During a typical campaign, the engine would fire every few minutes for periods of tens of seconds and the campaigns could last several months. It was important to be efficient and thorough in the configuration of the test, as it would take more than a week to make changes to the hardware inside. Therefore, the Ad Astra team had to develop very detailed test procedures and operations protocols in order to avoid any unnecessary chamber openings during experiment campaigns.

Despite the delay in delivering the magnet, the performance of the superconductor was definitely worth the wait. Full power on the VX-200 was achieved in September of 2009, a milestone that Dr. Chang Díaz remembers fondly, as he was in Boston, Massachusetts, attending a conference on space commercialization while the experiments in Houston were going on. While he participated in a business panel discussing the future of space commercialization, his team in Houston regularly updated him through his cell phone on progress on the full power experiment milestone they were attempting to achieve. Dr. Chang Díaz shared the anticipation with the audience as the calls continued to come in, updating

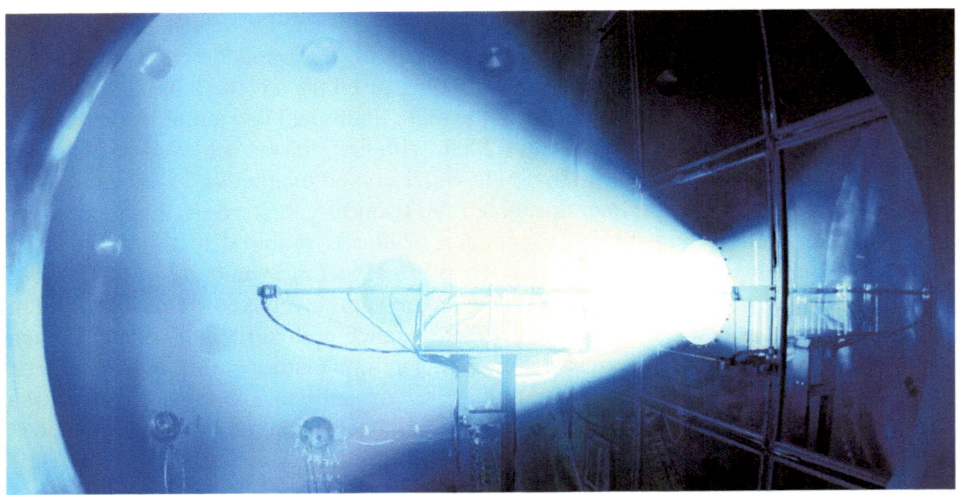

6.13 The 200kW VX-200 VASIMR® engine at full power fires with argon propellant in Ad Astra's 150 cubic meter vacuum chamber at the company's Webster, Texas facility. A movable instrumentations package measures the plasma performance over the plume cross section.

him every few minutes on the ever increasing power of the VASIMR® engine test. Finally, the much anticipated message of the achievement of the 200 kW arrived and Dr. Chang Díaz shared his excitement with the audience, who joined in applause.

REFERENCE

1. *Report of the President's Commission on Implementation of United States Space Exploration Policy*. E. C. "Pete" Aldridge Jr., Chairman, June 4, 2004.

7

The VX-200 and the Path to Commercialization

National Aeronautics and Space Administration
Office of the Administrator
Washington DC May 24, 2006

Dear Dr. Chang Díaz,

We have followed, with interest, the privatization of the Variable Specific Impulse Magnetoplasma Rocket (VASIMR®) project and are pleased with the success of our collaboration as defined in the Space Act Agreement signed June 26, 2005, between the National Aeronautics and Space Administration (NASA) and the Ad Astra Rocket Company. We have monitored your progress in the development of this technology with private funds and believe that your company's initiative could one day be applicable to NASA's exploration goals. We fully endorse such private efforts, and consistent with our Agency's needs, will consider using these capabilities when they come to fruition.

More specifically, we understand that one of your company's near-term objectives is the development of a 100 kilowatt VASIMR® engine, together with a space power and propulsion test facility, to be installed on the International Space Station (ISS). Such capability might be useful in providing drag compensation for the orbital complex or might evolve into a test bed for advanced electric plasma rockets in the future. NASA is interested in these concepts and willing to explore the feasibility of an addendum to the Space Act Agreement between NASA and your company. In order to determine if such an addendum is warranted, we should first collaboratively define the physical and functional interfaces required for the integration of such a system on the ISS, as well as your company's ability to sustain the development and operation phases.

In the long term, technologies such as VASIMR®, once successfully demonstrated, could play a role in the evolving space exploration architecture. I am directing William H. Gerstenmaier, Associate Administrator for Space Operations, to assign someone to work with your team on the potential for expanding our collaborative framework. I look forward to the continuation of a very productive relationship.

Sincerely
Michael D. Griffin
Administrator

© Springer International Publishing Switzerland 2017
F. Chang Díaz, E. Seedhouse, *To Mars and Beyond, Fast!*, Springer Praxis Books,
DOI 10.1007/978-3-319-22918-8_7

In the late spring of 2006, Dr. Chang Díaz received this letter from NASA Administrator Michael Griffin, which filled him and his company's leadership with optimism. The tone of the communication was very positive and the content, among other things, reiterated NASA's continued interest in Ad Astra's privately funded VASIMR® engine initiative and the agency's belief that the technology could be useful to NASA's future programs. More specifically, Dr. Griffin's letter indicated the possibility of conducting a test of Ad Astra's VASIMR® electric power and propulsion platform on the International Space Station (ISS) and acknowledged the value of one of Ad Astra's planned commercial uses of the technology; namely, to provide drag compensation to the ISS. In the letter, Dr. Griffin also informed Dr. Chang Díaz that he had assigned Associate Administrator for Space Operations, Mr. William Gerstenmaier to support a potential expansion of NASA's collaborative framework with Ad Astra.

While carefully measured in the wording of its content, the letter was a clear indication of interest at the top level of the agency and an implicit reaffirmation of the basic tenet in which the 2005 VASIMR® privatization was based. The letter also sent an important message to Ad Astra's investors, who in just one year had come forward with more than twice the funding that NASA had invested in 25. The fruits of their financial commitment were already evident in the rapid progress of the project, as VX-50 gave way to VX-100, which in turn paved the way for VX-200. NASA, too, had noticed that progress.

While electric propulsion within the agency remained in relative hibernation, Ad Astra's VASIMR® engine was making rapid advances in technology readiness level (TRL). At this pace, the company was quickly moving towards a space test of the technology, a near-term goal in its business plan, which would allow the firm to validate the immense volume of ground test data that had been gathered over several years of experiments. From his multiple flights into space, Dr. Chang Díaz had become convinced that, with ISS assembly nearing completion, a change in paradigm in electric propulsion technology development was now possible.

Ground test facilities were quite acceptable and economical for testing traditional low power electric thrusters at the few kW power level and below. High power electric rockets, on the other hand, placed high demands on a ground-based laboratory infrastructure, such as the requirement for large vacuum chambers with enough pumping throughput capacity to maintain an acceptable vacuum while the rocket is firing. The plasma physics of these devices is also strongly influenced by the presence of nearby chamber walls, which introduce complex electric and grounding effects – some unique to the particular thruster – that must be well understood and filtered out of the performance data. With high power systems, the demands levied by the rocket on the laboratory grow quickly and exponentially in both cost and complexity.

The ISS, on the other hand, offered a new and unique venue for testing electric thrusters, and more so with high power engines. The absence of chamber walls also eliminated complications in the performance measurements. Moreover, the ISS, itself an active spacecraft, would be the "ultimate thrust stand," providing an unequivocally clean measurement of thrust performance which would be ideal for comparing different propulsion technologies side by side. The US portion of the multinational orbital complex had just been declared a US "National Laboratory." It was already built and paid for and it was important to expand its use beyond biomedical research and microgravity and make full use of its unique capabilities.

NASA had grown accustomed to doing thousands of continuous hours of ground testing of electric rockets before a flight unit could be built. In an era of expensive single-flight robotic free-flyers, where early electric rockets could be flight-tested, it was logical to first secure their long-term reliability with extended test firings in laboratory test chambers. This approach also made sense from a cost point of view, as simple component failures and easily correctable early design weaknesses introduced unacceptable risk. These could irretrievably terminate an otherwise valuable space test, requiring the program to start again from the beginning.

This paradigm had begun to change in the 1980s with the evolution of human interactive experiments in the Shuttle program, but the complete shift occurred in the first decade of the 21st Century, when humans developed a permanent presence in space and astronaut training moved from task-based to skill-based. This transition led to a quantum leap in the sophistication of human-tended operations – both inside and outside of the spacecraft – that are possible in orbit. Suddenly, if the experimental hardware was properly built, it could be maintained, modified or otherwise upgraded in-situ by spacewalking astronauts, or in the pressurized shirtsleeve environment of a space station. Such human interaction allows a rational relaxation of the design requirements for the test article that, without sacrificing safety, could reduce the technology maturation cycle and hence the cost.

For the VASIMR® engine, besides taking advantage of space to verify ground performance data, the paradigm shift would enable space testing early in the technology maturation process, in order to "tease out" space-unique phenomena while design flexibility still existed and corrective action could be taken. For high power electric rockets, lengthy laboratory endurance tests *before* an early space "field" test could ultimately lead to a more lengthy and expensive development cycle.

Some debate the true scientific value of the present ISS, but no-one disputes that it stands as a marvel of human engineering, project management and international coordination and cooperation. To be sure, over the years, orbiting laboratories from Skylab to Mir to ISS, that grew out of the shadows of the Cold War, gradually evolved into instruments of international cooperation; outposts of biomedical research aimed at better understanding the effects of extended space missions on the human body. In a world increasingly rife with regional tension, they have also become beacons of hope for humanity's future and enclaves of national pride, as citizens from a growing number of "space faring" nations get a chance to orbit the Earth.

Beyond biomedical research, the potential of the ISS as a test bed for electric power and propulsion technologies, the two most critical elements required for a sustainable and robust human presence beyond LEO, is immense. Orbiting a few hundred kilometers overhead, with a permanent human presence and – notwithstanding the retirement of the Shuttles – an increasingly diverse array of resupply vehicles (from Russia, Europe, Japan and two US private firms), the ISS presents the first logical stepping stone for a VASIMR® space test. Over many years, from within NASA as the ASPL and now from outside NASA as the Ad Astra Rocket Company, the VASIMR® team has been in pursuit of such a demonstration, a test capability of value not just for VASIMR®, but for other high power electric rockets currently under development.

With this goal in mind, and energized by Dr. Griffin's letter, the Ad Astra team endeavored to pursue a new agreement with NASA to embark on such a test. To the Ad Astra team, it was abundantly clear from day one that finding a viable path through the complex bureaucratic maze of the ISS Program would not be easy. But paradigm shifts are *never* easy. To get started, there were some basic technical misconceptions on the part of the NASA management that needed to be addressed before moving ahead with a reasonable proposal. The first one was power.

Since the days of the ASPL, the VASIMR® researchers were often frustrated by off-the-cuff remarks – expressed in NASA technical meetings by VASIMR® detractors – referring to the VASIMR® engine as a "power hog," or with milder, but equally negative comments, such as "VASIMR® needs a lot of power," when comparing the technology to more conventional electric thrusters. To those unfamiliar with the technology, the instantaneous mental image elicited by those remarks was that of a heavy and inefficient system, dismissing the need to dig deeper and see the hard evidence to the contrary. It could, of course, not be discounted that such off-the-cuff remarks were precisely aimed at discrediting the engine as a viable propulsion technology. Over the entire journey of the VASIMR® engine development, the team had encountered its fair share of naysayers and detractors. These were expected and generally welcomed, as they helped the researchers focus their energies on answering many relevant technical questions. However, as with many other disruptive projects, a fair share of detractors are driven by non-technical agendas. As the ISS test discussions got underway, the subtle comments from some were seemingly aimed at extracting an early management dismissal of the ISS test, on the grounds that the station's power was insufficient to run VASIMR®.

The reality, of course, was quite different. Instead of *needing* a lot of power, the remarks could have been better worded as: VASIMR® *could process* a lot of power. Time and again, system studies consistently showed that, at high power, the VASIMR® engine's power specific mass – a parameter known as *Alpha*[1] (α) – was at or below that of its closest electric propulsion (EP) competitor at equivalent technology readiness. Nonetheless, the issue of ISS power availability in connection with a potential VASIMR® ISS test had to be addressed and the misconceptions clarified.

From the early concepts, generated and proposed at the ASPL for such a test on the ISS, the VASIMR® engine team knew full well that the ISS could not provide 200 kW of electric power to drive the engine. In fact, it was highly unlikely that more than 1-2 kW of continuous power would be available to support such experiments. The workaround for that constraint was simple and was proposed by the VASIMR® engine team well before Ad Astra's birth. The test article would include a battery pack, which could be trickle-charged at very low power over a couple of days in order to support continuous full power VASIMR® firings of a few minutes in duration. Repeatable test firings of a few minutes – a veritable eternity for a plasma engine – were all that was needed to verify the engine's performance. The plasma physics processes at work in the engine achieved steady state in a matter of a few milliseconds. As for thermal steady state equilibrium, which would take tens of minutes, the engine's thermal management could be easily controlled to retain more heat in between firings, thus achieving a thermal equilibrium by judicious throttling of the engine cooling system over many pulses.

---

[1] In expressing the Alpha (α) of the propulsion system, it is extremely important to account for the mass of the power processing unit (PPU), a detail which is often omitted in superficial high-level discussions.

Such nuances of the engine's operation were difficult to transmit to a NASA community of low power electric thruster advocates, who had grown accustomed to the more traditional approach of extremely long duration testing of electric thrusters in ground chambers prior to any space flight. Moreover, for Ad Astra, the objective of the VASIMR® engine space test was not to do long duration firings, but rather to achieve a thermal steady state in the engine components and measure the engine's performance in space. The space test data would be extremely useful to validate and further calibrate the ground test data. The unfortunate misconception over the power requirement, however, continued to be raised for years as a quick argument for dismissing the VASIMR® ISS test.

Other impediments to pursuing the VASIMR® test quickly arose that were more closely related to the ISS management structure. The transition of the ISS organization from a construction mode to a science operations mode was taking place slowly. This was completely reasonable for a project of such complexity and gargantuan proportions, being carried out in an unforgiving environment and with human lives hanging in the balance. The station was still being managed as a large operating spacecraft, with some room for science payloads. The general rules were that these payloads had to be simple, non-intrusive and fitting "standard" structural, power, telemetry and command interfaces. External payload sites were also sparse and mostly already taken up by other users, either for science or as temporary storage sites for equipment spares in anticipation of the Shuttle fleet's impending retirement. To be considered as an ISS payload, the VASIMR® had to "fit" within a very constrained accommodation envelope that all but eliminated its value.

Thus the VASIMR® test was immediately considered to be outside of the established ISS payloads envelope and hence it had to be considered as "a station element;" that is, part of the ISS vehicle. Such a designation had important economic implications for NASA. The prime contractor for the ISS vehicle, the Boeing Company, would have to get involved in the integration process and there seemed to be no budget allocation within the ISS Program to support something like that. Nonetheless, in 2009, the Boeing Company's Phantom Works, of Huntington Beach, CA, had approached Ad Astra about the potential of exploring a collaboration on the VASIMR® test. Under contract to DARPA (Defense Advanced Research Projects Agency), Boeing was developing the Fast Access Spacecraft Testbed (FAST), a lightweight and compact concentrator solar array scalable to hundreds of kilowatts. Combining FAST with VASIMR® propulsion could result in a ten-fold increase in spacecraft maneuvering capability over a chemically propelled vehicle of the same mass. Such capability was very much in line with Ad Astra's business horizon. The Ad Astra team saw this developing relationship as a potential incentive for NASA to move forward with the VASIMR® space test.

It took two years of informational meetings, technical briefings and assorted communications for Ad Astra to gather enough NASA support to move forward with an agreement for a VASIMR® space test. On September 9, 2008, Administrator Griffin organized a meeting in Washington with Dr. Chang Díaz and key NASA managers to reduce the bureaucratic viscosity and get the project moving forward. The participants included ISS Program Manager, Mr. Michael Suffredini, Associate Administrator for Space Operations, Mr. William Gerstenmaier and ISS Payloads Manager, Mr. William R. "Rod" Jones. In the meeting, Dr. Griffin directed his team to develop a Space Act Agreement (SAA) to pursue a VASIMR® test on the ISS. It was an election year in the United States, marking the end of President George W. Bush's second term and at NASA, there was concern that a new administration could imply a change of focus for the US Space Program.

The structure of the agreement was simple enough. Dr. Chang Díaz had proposed a simple schedule of technical "gates," to be defined by the parties, which both would have to cross as they moved to the space test. Both NASA and Ad Astra would work at their own expense, on their own side of the interface, to meet the requirements for each gate and achieve the integration of the VASIMR® test article on the ISS. There would be no exchange of funds and the parties would evaluate progress at each step of the way.

Four months later, the text of the agreement, skillfully guided through the Washington bureaucracy by Mr. Gerstenmaier's top aide, Mr. Jason Crusan, was complete, approved and awaiting signature. The US presidential election had also taken place in November and President-elect Barack Obama's transition team was already making the rounds through the various government agencies, including NASA. It was now clear that Administrator Griffin would not be shepherding the execution of the agreement after its signature. Fortunately, another able leader, Mr. William Gerstenmaier, would. He had taken an interest in the VASIMR® technology and, though peripherally at first, had observed the evolution of the project over many years at the JSC. Dr. Chang Díaz had known Bill since the 1980s. They had worked together on several Shuttle missions, some of which Dr. Chang Díaz had flown and others which he had supported, alongside Bill, in Mission Control as a Capsule Communicator (CAPCOM). In his early NASA career at the JSC, Bill Gerstenmaier had distinguished himself as an outstanding flight controller. Over the years since the two had worked together, Bill had risen through the NASA ranks to positions of increasing leadership and responsibility, which quickly blended his skills as a consummate engineer with those of an excellent manager.

On December 8, 2008, William Gerstenmaier, for NASA, and Franklin Chang Díaz, for Ad Astra Rocket Company, signed the agreement entitled:

NON-REIMBURSABLE SPACE ACT AGREEMENT

BETWEEN

AD ASTRA ROCKET COMPANY

AND

THE NATIONAL AERONAUTICS AND SPACE ADMINISTRATION

FOR

DEMONSTRATION OF

THE VARIABLE SPECIFIC IMPULSE MAGNETOPLASMA ROCKET (VASIMR™)

ABOARD THE INTERNATIONAL SPACE STATION

The following excerpts from the agreement illuminate the spirit of collaboration that had engendered it and a recognition of the paradigm shift in the utilization of the ISS as a national laboratory:

"The purpose of this Agreement is to facilitate and conduct a space flight test of a VASIMR™ engine on the ISS (the "Project"). The Parties hereby agree to embark on a series of sequential phased activities to achieve this objective."

"…NASA plans to operate a share of the U.S. accommodation on the ISS as a National Laboratory in accordance with the NASA Authorization Act of 2005, Section 507 (P.L. 109-155). This Project with the VASIMR™ test serves as a "pathfinder" for the ISS National Laboratory by demonstrating a new group of larger, more complex classes of science and technology payloads, encouraging others to pursue similar projects and facilitate their efforts by providing a model for implementation. These larger, more complex classes of payloads require a larger effort by NASA to fully integrate them with the ISS and this multi-gate approach will allow NASA to assess the requirements on an incremental basis while proceeding to flight."

"Discussions between NASA and AD ASTRA, aimed at a VASIMR™ ISS demonstration have been ongoing since 2005 under previous agreements. A number of joint NASA/AD ASTRA studies and assessments have been conducted, which support the technical feasibility of such a test and have led to the present concept of a 200 kW VASIMR™ package, operated in short bursts of up to several minutes from a battery pack periodically charged by the ISS power bus."

FROM ROCKET SCIENCE TO FINANCIAL INNOVATION

The signature of the agreement for the ISS space test energized the Ad Astra team, which was recovering from two major non-technical setbacks. One was Hurricane Ike, a Category 4 storm that devastated much of the coastal areas around Houston in mid-September of 2008. The sustained gale force winds and flooding had inflicted major damage to the Ad Astra facilities and the homes of several of the employees. Recovery efforts lasted well into 2009. Fortunately, a number of non-export-sensitive supporting projects and some of the Houston personnel were temporarily detailed to the company's Costa Rica subsidiary, which had already developed a fully operational plasma laboratory.

The second setback was the US economic crash of 2008, which resulted in a slowdown of investment and affected Ad Astra, along with thousands of other US and international companies. Since its inception in 2005, Ad Astra had exercised not only technical but also financial innovation and the company had begun to test novel concepts for raising capital, particularly in Dr. Chang Díaz's home country of Costa Rica where the company was quickly gaining strong popularity and media visibility. In April of 2008, Ad Astra inaugurated – and became the first company to enter – a new experimental private stock market named "Mercado Alternativo Para Acciones" (MAPA), nested inside Costa Rica's Stock Exchange, known as the "Bolsa Nacional de Valores." MAPA was modeled after the London Stock Exchange's Alternative Investment Market (AIM), a private investment infrastructure specially designed for developing high-risk companies seeking growth investment. A step before a public offering, AIM helps strengthen corporate governance with regulation appropriate to smaller firms.

Ad Astra had raised several million dollars in the form of private placements in Costa Rica, but by 2008 it had quickly topped the 50 investor limit allowed by Costa Rica's regulation for a private company. While the transition to MAPA did not change that number, it served to prepare the company for the next step it was contemplating: a potential public offering of its common shares in the Costa Rican Stock Exchange. Such a move had not been attempted by an American company before, but was allowed under Regulation S of the US Securities and Exchange Commission (SEC). Specifically, the rule allowed a private US company to list its shares in a foreign market, as long as the purchase of shares was only available to non-US persons. In this way, Ad Astra could remain a private company in the US but become a publicly held company in Costa Rica.

In 2010, Ad Astra made history again in Costa Rica's financial community, as it exited MAPA and entered the full Costa Rican market as a publicly traded company under the country's first Restricted Public Offer (RPO). The RPO was a new designation developed by the Costa Rican government to allow high risk companies, such as Ad Astra, to benefit from the liquidity of a public market but only in a restricted way. The purchase of Ad Astra shares was available only to "sophisticated" investors. These were individuals or institutions of sufficient net-worth to understand the financial risk in the investment and to be able to sustain a loss of their capital if the company were to fail.

7.1 In April 2008, Ad Astra Rocket Company becomes the first firm to enter the Costa Rican Mercado Alternativo para Acciones (MAPA) a private market, nested within the Costa Rican Stock Exchange and modeled after London's Alternative Investment Market (AIM). (L-R): Oscar Luis Chaves, President Aldesa Investment Bank, Dr. Chang Díaz, CEO, Ad Astra Rocket Company and Orlando Soto, President of the Board of Directors of the Costa Rica Stock Exchange, sign the entry documents in San José, Costa Rica.

As a highly educated and forward thinking nation, Costa Rica was one of the fastest growing economies in Latin America prior to the 2008 market crash. When Ad Astra entered MAPA, this nation of strong social, economic and political stability did not hesitate to join the entrepreneurial initiative of one of its most beloved citizens and many Costa Rican sophisticated investors joined their US counterparts as Ad Astra shareholders. The market slowdown, however, dampened some of this investment and induced Ad Astra to implement additional revenue-generating strategies. These have strengthened the company through diversification into renewable energy technologies, including water electrolysis for hydrogen production from wind and solar power, of great relevance to Costa Rica.

Ad Astra's diversification into the renewable energy field drove it to establish two other subsidiaries, Ad Astra Energy and Environmental Services in the US and Ad Astra Servicios Energéticos y Ambientales SRL (AASEA) in Costa Rica. The company believes that a nation like Costa Rica, which has no oil or natural gas resources but is rich in hydroelectric, geothermal, solar and wind resources, could transform its economy to the hydrogen cycle, utilizing this medium as a large, clean energy storage through water electrolysis. The country already produces more than 90 percent of its electricity from clean renewable sources. Transportation, however, remains the largest energy demand, consuming more than 100,000 TJ (terajoules) of energy per year. Unfortunately, the transportation infrastructure is largely based on imported petroleum products and its carbon footprint is very high. Since 2009, the company has been developing a solar- and wind-based hydrogen transportation ecosystem and, in 2013, deployed the first hydrogen production and storage facility in Central America utilizing water electrolysis. The company is also applying hydrogen technology to improve the efficiency of biofuels, such as bio-digester gas in rural areas. In April of 2016, the Costa Rica subsidiary became 100 percent solar and now generates a significant amount of excess energy, which is currently being stored in the national grid under an agreement with the country's electric utility. In mid 2017, the company, in collaboration with the Costa Rican government and several other companies in Costa Rica, the US and Europe, plans to inaugurate the first hydrogen fuel cell urban transportation ecosystem in Central America in the city of Liberia, capital of the province of Guanacaste.

PROBING THE VX-200™ PERFORMANCE ENVELOPE

Ad Astra's diversification into the renewable energy field generated important revenue for the company which helped it weather the lean years following the financial crash, but the company's main line of activity, the development of the VASIMR® engine, continued steadily, primarily in Houston. In 2009, following the recovery from Hurricane Ike and the delivery and integration of the VX-200™ superconducting magnet, Ad Astra initiated a series of activities aimed at keeping pace with its responsibilities pursuant to the NASA ISS agreement. These included a number of experimental campaigns to increase the TRL of the VASIMR® engine and demonstrate its competitive performance. Despite the slowdown in investment, the pace of the technology maturation remained brisk, as the new VX-200™ prototype was now fully functional, though the complete operational envelope had not yet been explored.

An important test had also taken place in late 2007 at the University of Michigan's Electric Propulsion Laboratory, which validated the effectiveness of Ad Astra's Plasma Momentum Force Sensor (PMFS) to measure the thrust of the VASIMR® engine. The PMFS is a thrust measurement instrument, developed by the VASIMR® team in the later years of the ASPL, to obtain the thrust of an electric rocket by direct measurement of the momentum in the plasma plume. For high power electric rockets, this method is far less expensive than the more traditional "thrust stand," which would require the entire rocket to be suspended on a very sensitive structural balance. To prove the equivalence of the PMFS and "thrust stand" measurements, a "blind test" was conducted in Michigan, a third party facility, using both methods on a known Hall Effect Thruster. The results, published in May of 2009 in the peer-reviewed *Journal of Propulsion and Power*, showed that the two methods produced the same result.

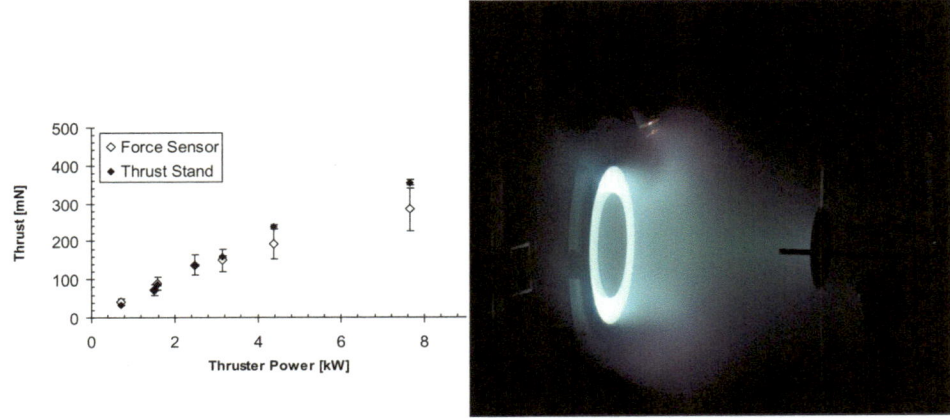

7.2 In October of 2007, a "blind" thrust measurement test of Ad Astra's thrust target at the University of Michigan Electric Propulsion Laboratory successfully predicted the measurement using a traditional thrust stand, thereby validating the effectiveness of Ad Astra's method.

Several highly successful experimental campaigns followed between 2009 and the end of 2012 that fully explored the VASIMR® operational envelope. These campaigns also achieved advances in system control, particularly with regard to the complex physics of the start-up sequence. This process was erratic and somewhat unstable at first, but later was successfully controlled with a deeper understanding of the physics at play during the transient and start-up phases in these high density RF plasma discharges. For example, in the early phases of the start-up, the impedance of the plasma is very high, as most of the target gas is still in its neutral state. However, as the discharge proceeds in the scale of tens of microseconds, the plasma becomes more conducting and its impedance drops significantly. Despite these dynamics, the proper impedance matching conditions must exist throughout the start-up. This required considerable optimization and proper design.

Focused studies also demonstrated the engine's high efficiency and the thrust/I_{sp} variability at constant power; the VASIMR®-specific feature that begets its name and distinguishes it from other electric thrusters. In early 2013, experiments with krypton propellant were initiated to explore further expansion of the engine's operational envelope to include other fuels. In this way, the operational envelope of the engine continued to be explored and characterized and the results continued to follow, validating the theoretical predictions which had been generated by the VASIMR® team back in the ASPL days.

The results of the VX-200™ experimental campaigns were quite compelling and showed the progress the team was achieving over several years of effort. The thruster efficiency as a function of the specific impulse was a critical metric. Efficiencies greater than 50 percent were highly desirable. As the first campaign got underway in the fall of 2009, it was clear that much remained to be learned about the rocket's optimal operation. These early experiments showed the expected efficiency increase with second stage power but fell short of the desired 50 percent. A second campaign initiated in the spring of 2010 showed a dramatic improvement of the performance, brought about by advances in the field profile and fueling control techniques. These experiments were extended in a third campaign in 2010, where efficiencies in excess of 70 percent were reached. Further experiments, spanning 2012 and early 2013, fully optimized the engine's operation with argon propellant and began to explore the utilization of heavier propellants such as krypton.

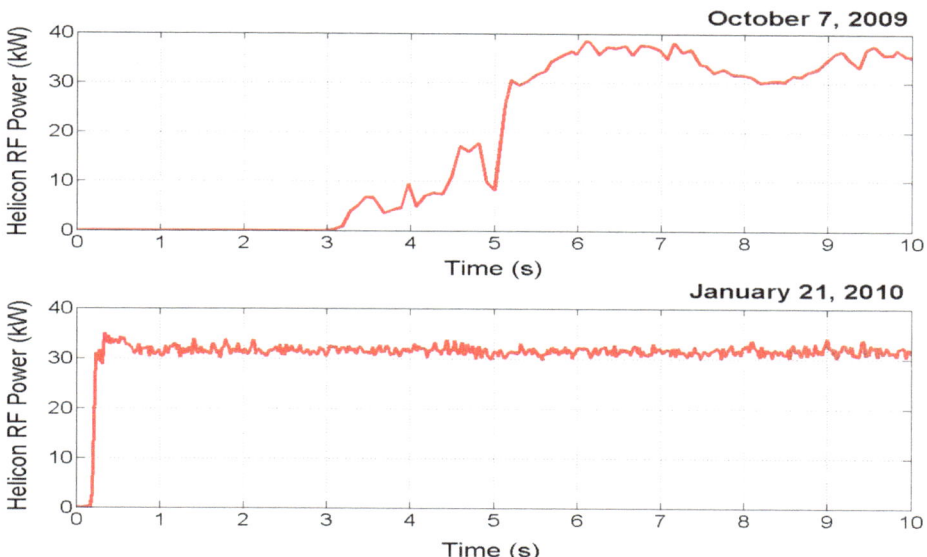

7.3 Start-up control was achieved after the first high-power experimental campaign in 2009, when the team gained a deeper understanding of the complex plasma physics at play in the high-density discharge.

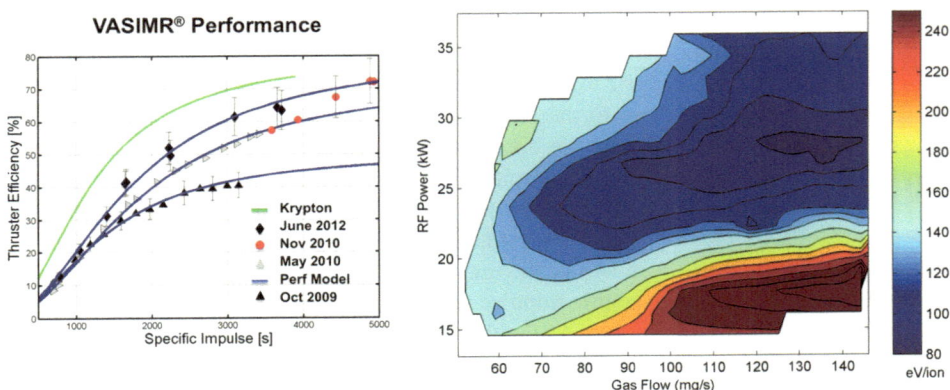

7.4 Four experimental campaigns (left) from 2009 to 2013 mapped the operational envelope of the VASIMR® and demonstrated the engine's high (>70 percent) efficiency. Further reductions in the ionization cost (right) to below 100 eV/ion were achieved.

In addition to engine performance and further reductions in ionization cost to levels below 100 eV/ion, another important objective was achieved during the 2010-2013 experiment campaigns; namely, the demonstration of plasma detachment from the magnetic nozzle. This important issue had lingered without conclusive experimental verification since the early NASA days at the ASPL and had been a very contentious subject during the 2002 VASIMR® peer review. As had been indicated to the reviewers, the VX-50 apparatus and the small vacuum chamber existing at that time precluded that observation. The verification had to wait for the more suitable experimental configuration provided by the VX-200™, operating in Ad Astra's large vacuum chamber. Ad Astra's Christopher Olsen, a company research scientist and a PhD Candidate at Rice University, obtained the first conclusive measurements of the detachment process and described the phenomenon as driven by a "loss of adiabaticity" in the plasma plume downstream, where the magnetic field's grip on the plasma was insufficient to prevent it from breaking free. His observations were published in 2013 in the peer-reviewed journal *Transactions in Plasma Science*.

Adiabaticity in a plasma is a way of describing relatively slow changes in time and space as one examines the dynamics of individual charged particles in the plasma. For example, if the motion is adiabatic, as the particle corkscrews about the magnetic field line along which it is moving, the magnetic field strength changes little over each orbit. In that way, the particle is able to follow the field line and if the field line bends slowly, the particle is able to "keep up." However, if the field line direction changes significantly over one particle orbit, the particle "loses track" and is no longer guided by it. The motion is said to be non-adiabatic and the particle is no longer confined by the field. An analogy can be used to describe the loss of adiabaticity in a more mundane, if not familiar, process. If a reckless, fast moving driver on a slippery highway waits too long to take the exit, his sudden turn leads to a loss of adiabaticity – and a potential crash – as the tires of his vehicle lose grip on the road.

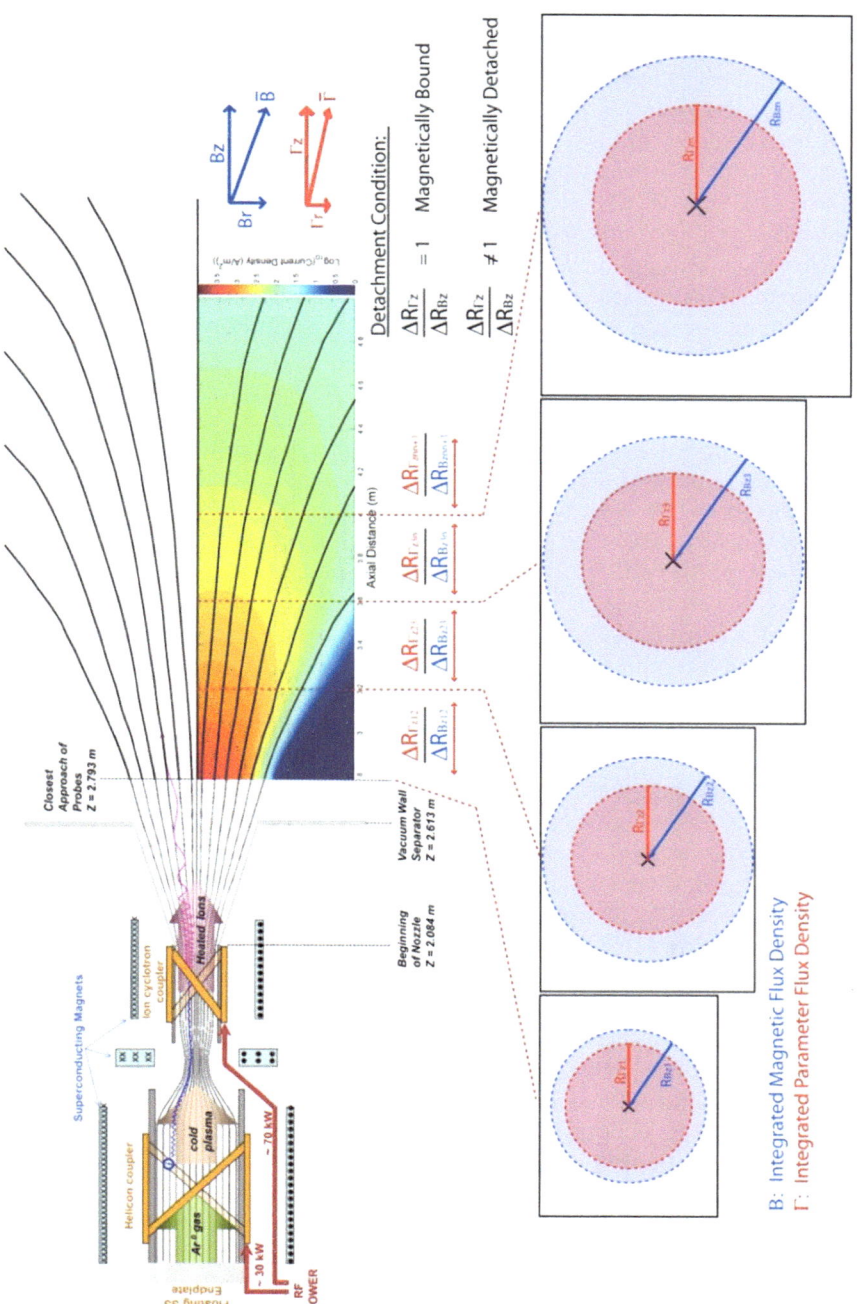

7.5 In the VASIMR® magnetic nozzle, the magnetic flux (blue) expands faster than the ion flux (red), indicating the detachment of the latter from the field.

THE ROCKY ROAD TO THE ISS

Following the signature of the SAA in December of 2008, interaction with the ISS Payload Integration Office began in early 2009. The first goal for the parties was to understand the unique classification of the VASIMR® test article and identify a potential location on the station where it could be attached. The VASIMR® flight engine itself was to be part of an integrated Ad Astra Power and Propulsion Test Platform, called "*Aurora*," that would provide structural support, as well as battery energy storage, thermal control, telemetry and command capability, from onboard the ISS as well as from the ground. The entire 4.5-ton package was expected to be delivered to the ISS by one of the approved NASA commercial carriers. No down selection on the carrier had yet been made, but the technical team was leaning towards Orbital Science Corporation's (now Orbital-ATK) Cygnus transport vehicle as the most suitable venue.

Unfortunately, on the NASA side, the integration process quickly became bogged down in the complex ISS management structure. While the agreement to support the test had been signed, the space agency's financial commitment was simply not there. Moreover, despite positive peer reviews and compelling progress on the technology maturation, the project's detractors had not gone away completely. On June 9, 2009, as required by the ISS Agreement, Dr. Chang Díaz and NASA's Payload Integration Manager, Mr. William R. Jones, signed the VASIMR® Payload Integration Agreement, thereby crossing the first identified gate. However, the agency could not find the internal funds necessary to support Mr. Winn Reed, a Boeing employee who had been assigned to coordinate the integration process as Payload Integration Manager (PIM).

Other indicators appeared to forecast additional NASA headwinds for the project, as Ad Astra's formal proposal for an "Electric Propulsion Test Platform ISS National Laboratory Pathfinder Mission," submitted in February of 2010, was never answered. Ad Astra viewed the proposal as a required response to the ISS National Laboratory Announcement of Opportunity NNH09CAO003O, which was released by NASA on August 6, 2009 and was itself a stated requirement of the SAA.

In July of that year, General (USMC, ret) Charles F. Bolden Jr., a former astronaut, and Ms. Lori Beth Garver, a Senior Space Policy Advisor to President Obama, had been sworn in as NASA Administrator and Deputy Administrator respectively and were starting to get the pulse of the agency. Their arrival had given the VASIMR® team much hope of a renewed NASA commitment to the spirit of the ISS Space Act Agreement. During the congressional hearings for his nomination as NASA Administrator, Charlie Bolden had publicly praised Dr. Chang Díaz's work on the VASIMR® engine and had expressed its potential for NASA. Charlie Bolden, however, was a former astronaut colleague and friend of Dr. Chang Díaz and, as NASA Administrator, his support for the VASIMR® engine project fell under intense scrutiny, a condition that actually increased the bureaucratic viscosity of the ISS test and the Ad Astra – NASA relationship.

During this period, Ad Astra was also focusing its efforts on the recovery from Hurricane Ike and the experimental campaigns with the VX-200™. The company also began structuring an internal documentation process for the *Aurora* project and developing concepts of operations (CONOPS). Several important milestones were accomplished in 2010 towards this goal, including the development of a conceptual design for the *Aurora*

Platform, product and system requirements documents and a System Engineering Management Plan. All of these were completed in anticipation of a re-energized integration activity with NASA.

By early 2010, however, mixed and confusing signals from NASA in Washington regarding its commitment to stay true to the spirit of the agreement prompted Ad Astra to seek new avenues of NASA collaboration at the local level, through its active 2007 Umbrella Space Act Agreement with the JSC. The agreement was an open vehicle for JSC and Ad Astra to work together on the VASIMR® technology maturation, and several areas of common interest relating to the ISS test had come into focus. The Astronaut Office at JSC had also remained loosely connected to the project and astronauts Lt. Cdr. (US Navy), David Bowen and US Navy physician, Dr. Lee Morin, would come to Ad Astra from time to time to keep abreast of progress. David was a submarine officer in the US Nuclear Navy and Dr. Chang Díaz had expressed to him his interest in the Navy's operational ships, driven by nuclear-electric propulsion, as good analogues for his ultimate human interplanetary spacecraft driven by multi-megawatt nuclear-electric VASIMR® propulsion. Lee, on the other hand, a family physician by training, had acquired great skill in computer architectures and digital human interfaces. He had been interested in addressing Ad Astra's requirement to control the VASIMR® test engine from on-board the station. This was an important design feature, involving conceptual human-computer interfaces with the required software that could be explored early.

Other astronauts who had moved to management within the JSC structure had also been supportive from their new positions, including Dr. Janet Kavandi, who had become Director of Flight Crew Operations, and Col. (US Army, Ret.) William McArthur, who became Director of JSC Safety. There was also support at the international level, from Canadian veteran astronaut Chris Hadfield, who was preparing to command the ISS on his next flight, as well as two of his new astronaut colleagues, Dr. David Saint Jacques, a physician, and Col. Jeremy Hansen, a Canadian Navy pilot, all of whom promoted the VASIMR® technology in their home country. One of the critical components of the VASIMR® engine, the RF subsystem, incorporates high power solid state RF generators initially provided by Nautel Ltd., a Canadian company.

On March 12, 2010, Mr. Steve Altemus, then Director of Engineering at the JSC, invited Dr. Chang Díaz and Dr. Mark Carter, Ad Astra's Senior Vice President for Technology Development, to a meeting with his top management team. Mr. Altemus was a strong supporter of the VASIMR® engine project and wanted his organization to provide assistance in the ISS integration of Aurora. The meeting was of great importance towards initiating a comprehensive technical study of the ISS test and examining all the potential integration issues, many of which were often discussed in briefings to NASA top managers lacking adequate supporting information.

To coordinate the effort, Dr. Chang Díaz requested that Mr. Altemus assign a representative from his team as point of contact and proposed two individuals who were thoroughly familiar with the VASIMR® engine; Mr. Kenneth Bollweg and Mr. Trent Martin. Both of them had worked with Dr. Chang Díaz in the past and had a great deal of experience in the integration of complex Shuttle and ISS payloads, including the Alpha Magnetic Spectrometer (AMS) that Dr. Chang Díaz had operated on STS-91 and which was to be flown to the ISS for a second space mission.

Unfortunately, both Mr. Bollweg and Mr. Martin were busy supporting other major projects and would not have enough time to provide a steady commitment. As an alternative, Mr. Altemus proposed Dr. Harold "Sonny" White, an engineer with the robotics branch who had expressed an interest in electric propulsion. Dr. Chang Díaz knew Sonny from earlier years, as the former had visited Dr. Chang Díaz at the ASPL. Dr. White had a personal interest in exotic and futuristic space propulsion concepts and had come to discuss his latest idea for tapping energy from the quantum vacuum as a means of propelling a spacecraft.

Though Dr. Chang Díaz appreciated Sonny's interests, he was not completely sure that these aligned well with the VASIMR® project. Nonetheless, he was thankful for the support of Mr. Altemus and decided to give it a try to move the collaboration forward. Sonny White was assigned to support the Ad Astra collaboration in April of 2010. He tackled his new assignment with a great deal of management energy and enthusiasm. He organized and chaired a new team called the VASIMR® Engineering Integration Working Group (VEIWG), with bi-weekly teleconferences from Ad Astra in Houston. The VEIWG grew very quickly into an unwieldy and time-consuming activity, with the rapid accretion of dozens of representatives from virtually all NASA centers. For the relatively small Ad Astra team with its highly focused technology maturation program, the massive two hour bi-weekly discussions became too distracting and overwhelming, and ultimately of no practical use to the project. After a year of meetings, the process was discontinued.

Two significant products, however, resulted from the VEIWG deliberations over most of 2010: first, an internal NASA proposal to develop the VASIMR® Electric Power and Propulsion Test Platform as a multicenter effort with Ad Astra; and second, a comprehensive study for the Office of the Chief Technologist that assessed the integration of the VASIMR® engine on the ISS. The proposal was in response to a NASA-wide call for advanced technology initiatives. It was a team effort, involving the Johnson and Marshall Space Flight Centers, the Glenn Research Center and the Ad Astra Rocket Company. The proposal was highly successful, as it became one of the finalists of a long process of evaluation, competing with dozens of other proposals from all the NASA centers. In the end, however, the program was cancelled.

The second significant product came from another segment of Mr. Altemus's organization that became active within the framework of the VEIWG. It was known as the ISS Vehicle Integrated Performance and Resources (VIPeR) team and consisted of a group of NASA and Boeing engineers who specialized in analyzing unique ISS integration challenges. The team was led by Dr. Jack Bacon, an expert in spacecraft integration with a great deal of experience in large spacecraft issues. Jack was also familiar with the plasma physics of the VASIMR® engine and could readily distinguish the real VASIMR® ISS integration challenges from those which were manufactured concerns based on anecdotal and/or erroneous facts.

The Viper team submitted their findings on March 24, 2011 in a detailed briefing, which was presented to ISS Program Manager, Mr. Michael Suffredini, and later to the Office of the Chief Technologist in Washington. Their main conclusion was:

> "As a result of the VEIWG efforts, the technical community feels that there are no show-stoppers for the AURORA to be successfully integrated and safely operated on board the ISS."

Despite these conclusions, no funding to support the agency's obligations under the agreement became available and gradually the activities of the VEIWG were de-scoped and eventually ceased. It appeared that the only plausible "show stopper" was the agency's own inability to secure funding for the test. Two agency organizations seemed to clash on the importance of its pursuit: The Space Technology Mission Directorate, who favored continued work on the Hall thruster, a low power electric propulsion technology being developed by the NASA Glenn Research Center; and the Space Operations Mission Directorate, the original architects and signatories of the SAA. Within this apparent struggle, however, there seemed to be enough interest to continue to explore funding avenues, and the term of the SAA, which was due to expire in December of 2011, was extended for two more years. Moreover, another collaborative agreement was signed, this one between Ad Astra and the ISS safety organization of the JSC Safety Directorate. This agreement was to provide support and guidance to Ad Astra on navigating the safety requirements that NASA would levy on the VASIMR® ISS test.

Despite such bewildering signals, Ad Astra continued with the VASIMR® technology maturation and its portion of the ISS integration responsibilities. In 2010, a high fidelity engineering mockup for the *Aurora* Electric Power and Propulsion Test Platform was initiated in Costa Rica, as an Ad Astra-led project, with a team of six local advanced manufacturing companies. The project involved the high fidelity design, construction and test of a potential full-scale *Aurora* ISS platform mockup that could support the ISS testing of a 200 kW VASIMR® flight engine. The project was sponsored by the Costa Rican Science and Technology Research Council, an arm of the Costa Rican Ministry of Science and Technology, and was successfully completed in 2011. Also in that year, in Ad Astra's Houston facility, the development of a full-scale mockup of the VF-200™ -1 VASIMR® flight engine was initiated, which was completed in 2012. Both of these mockups were considered important architectural pathfinders, exploring launch stowage configurations and EVA servicing techniques.

For Ad Astra and the VASIMR® team, the years following the NASA privatization transition were extremely productive, as the infusion of fresh capital enabled major advances in the technology. The intellectual property associated with those advances was captured on November 26, 2013, in a new Ad Astra owned patent (number 8593064 B2), entitled "*Plasma Source Improved with an RF Coupling System*." Other new features of the rocket design, which also constitute advances from the original NASA patent, will remain as Ad Astra trade secrets.

In early 2010, with the VASIMR® technology approaching space test maturity, the company began focusing on developing an array of potential near-term commercial applications in high power solar-electric propulsion. Human missions to Mars and points in deep space using VASIMR® nuclear-electric propulsion would have to take a backseat to the more immediate, revenue-generating, high power solar-electric applications. The US National Research Council had identified high power electric propulsion, in the range of 50 to 600 kW, as a critical technology for NASA and the nation. NASA had accepted those findings and began addressing the technology gaps. Unfortunately, although traditional low-power gridded ion engines and Hall Effect Thrusters were already operational and demonstrating tremendous value to missions such as Deep Space-1 and Dawn, as well as providing station-keeping propulsion to a growing number of commercial satellites, the technology had not matured significantly beyond 5 kW. An advanced 12 kW Hall thruster was under development by NASA's Glenn Research Center, with the expectation that clusters of these engines, working together, could enter the high power regime. The Ad Astra team recognized the strong momentum these traditional technologies enjoyed within NASA, but knew that at their current pace of development, VASIMR® engines could also be well poised to be worthy competitive contenders in the high power niche.

In 2013, the company moved to implement further enhancements in the VX-200™ design, with the adoption of new materials and manufacturing techniques to make the engine more compact and lightweight. The relationship with NASA also strengthened. In the fall of 2014, NASA announced its Next Space Technology Exploration Partnerships (NextSTEP) Broad Agency Announcement, a call soliciting proposals from US private industry for high power electric propulsion. Its objective was to bring medium TRL technologies in this field to TRL 5. Ad Astra's proposal was selected for funding in March of 2015. Contract negotiations were completed in August of 2015, with the signature of a three-year, fixed price agreement for a total value of just over $9 million. Under this award, Ad Astra will conduct a long duration, high power test of an upgraded version of the VX-200™ VASIMR® prototype, the VX-200SS™ (for steady state), for a minimum of 100 hours continuously at a power level of 100 kW. These experiments will demonstrate the engine's new proprietary core design and thermal control subsystem and provide better estimates of component lifetime. The tests will be conducted in Ad Astra's large, state-of-the-art vacuum chamber in the company's Texas facility.

On July 19, 2016, Astra successfully completed all milestones and deliverables for the first of its three-year NextSTEP contract with NASA. After a year-one performance review that took place at NASA Headquarters in Washington DC, Ad Astra received NASA's approval to proceed with year-two activities. The test article is the VX-200SS™, which includes a new proprietary rocket core design and will be capable of operating indefinitely in a thermally stable mode under space-like vacuum conditions.

Important achievements for year-one included the redesign and manufacturing of the new VX-200SS™ rocket core and new vacuum and thermal management subsystems for the laboratory. The last two are needed for handling the vacuum requirements and the high thermal load arising from the rocket's three-million-degree plasma exhaust. Other milestones included the refurbishment of the rocket's cryogen-free superconducting magnet and the high power RF generators. By early 2017, Ad Astra had successfully completed the first set of full-power plasma tests on the VX-200SS™. A new campaign of high power tests is due to begin in mid 2017. These tests will continue through the remainder of 2017 and into early 2018, at increasingly longer pulse lengths. Ad Astra expects to reach the 100 hr./100 kW goal by mid-2018. The planned long duration tests with NASA aim to demonstrate the VASIMR® engine's durability and thermal control.

Ad Astra views a successful 100-hour test as the last major ground milestone before the VASIMR® engine is ready for a test flight in space. While the ISS venue remains an option, the company continues to explore other avenues which may become available in the near future, including a robotic free-flyer. Much has changed in the world's space activities since the early 1980s and opportunities not considered viable then are viable now. After nearly four decades of development, from an early physics concept to near spaceflight readiness, the VASIMR® engine's journey to space is about to begin.

8

A Bridge to the Future

For a private rocket company, powering the ships that may someday send humans deep into space is a lofty goal. Yet to the average investor, more concerned with internal rate of return (IRR) and gross margin, such romanticism, while laudable, is financially unsellable. For Ad Astra Rocket Company, the practicality of near-term, revenue-generating commercial applications of the VASIMR® engine quickly trumped the long-term ambition of ultimately supporting the "Nautilus Paradigm." Fortunately, those applications became clear and evident as the company studied the market trends and future needs of commercial space. High power electric propulsion, in the range of hundreds of kilowatts, provides clear business advantages in the maintenance and servicing of the near-Earth satellite infrastructure. It also became clear that rapid advances in solar-electric space power would enable high power solar-electric VASIMR® robotic missions that could support a very strong, near-term business model; one enabling significantly more capable space logistics missions than those provided by chemical propulsion.

The company identified four main near- to medium-term niches for VASIMR® engines driven by solar-electric power: 1) in-space logistics, which addresses the movement and servicing of payloads in space, including satellite refueling, maintenance and repair, repositioning and end-of-life disposal, 2) orbital re-boost and/or atmospheric drag compensation of large space stations and platforms in low Earth orbit (LEO), 3) in-space resource utilization, including the mining of asteroids and comets, and 4) a solar-electric deep space catapult, delivering fast scientific payloads far into the outer solar system.

The fourth niche, described later in this chapter, would enable rapid and economical exploration of the icy moons of Jupiter and Saturn. Perhaps even more than Mars, these bodies have recently captured the attention of the scientific community with tantalizing evidence of a potential life-supporting ocean environment beneath their icy surfaces. Moreover, as a derivative of the third application, and assuming sufficient power availability and early detection, the VASIMR® engine could also be used as a planetary defense tool, to deflect an incoming medium-size asteroid threatening a collision with the home planet.

© Springer International Publishing Switzerland 2017
F. Chang Díaz, E. Seedhouse, *To Mars and Beyond, Fast!*, Springer Praxis Books,
DOI 10.1007/978-3-319-22918-8_8

8.1 Powered by a 200 kW VASIMR® engine, Ad Astra's solar-electric LEO Space Cleaner approaches a spent Zenit upper stage for capture and disposal.

THE VASIMR® ORBITAL SWEEPER

A major part of the first business niche addresses mounting concerns over space debris. This issue has begun to attract a great deal of attention, as the accumulation of more than 500,000 Earth orbiting objects continues to grow. The problem of orbital debris was immortalized in 2013, in the award winning sci-fi movie *Gravity*. The movie illustrates the destructive effect of the so-called "Kessler Syndrome[1]," a chain reaction that precipitates when collisions between satellites and their resulting debris grow out of control. The trigger incident in *Gravity* occurs when Russia launches a missile to destroy one of its own satellites. This action unleashes a hyper-velocity cloud of space debris that destroys other satellites in LEO. Zipping around the planet every 90 minutes, the cloud grows as it destroys everything in its path, unfortunately including a Space Shuttle that happens to be in LEO at the time. The lethal cloud destroys the ship and kills most of the crew.

While Hollywood's technical accuracy in *Gravity* may have been questionable, the scientific fundamentals of the problem it describes are accurate. Orbital debris and its threat to orbiting spacecraft is a significant concern for satellite operators, astronauts and

---

[1] The Kessler syndrome is named after Donald Kessler, a NASA astrophysicist who predicted a potential runaway condition of debris causing further debris as the satellite population increased and collisions were more frequent.

mission controllers. Under the right conditions, the Kessler Syndrome could persist long enough to render the space in LEO virtually uninhabitable for an extended period. Indeed, orbital debris triggers have already occurred. In 2007, in a demonstration of its anti-satellite missile capability, China destroyed FY-1C, one of its weather satellites in polar orbit. The resulting explosion, at an altitude of 865 kilometers, created thousands of hypervelocity fragments that are likely to remain in orbit for decades. Two years later, an active US Iridium satellite collided with Kosmos 2251, a defunct Russian satellite, generating an estimated 2,000 additional pieces of space debris.

The Iridium-Kosmos collision was a watershed event that brought to light the fact that much more work is needed to ensure the safe separation of space objects. With more than 500,000 orbiting pieces being tracked by the US Space Command, the control systems currently ensuring the safe separation of orbital assets are becoming overloaded. Moreover, the risk to humans is also a growing concern. During the Shuttle era, several collision avoidance maneuvers were conducted by the orbiters (about four per year) to avoid space debris. In 133 Shuttle missions, the orbiters were struck by orbital debris 1894 times[2], damaging windows, radiators and other external surfaces.

Fortunately, the Kessler Syndrome is a decades-long process, giving us sufficient time to develop a practical method to avoid it. Nonetheless, we must not delay in addressing the threat, as a single impact could be lethal to astronauts working in orbit. It is important to focus

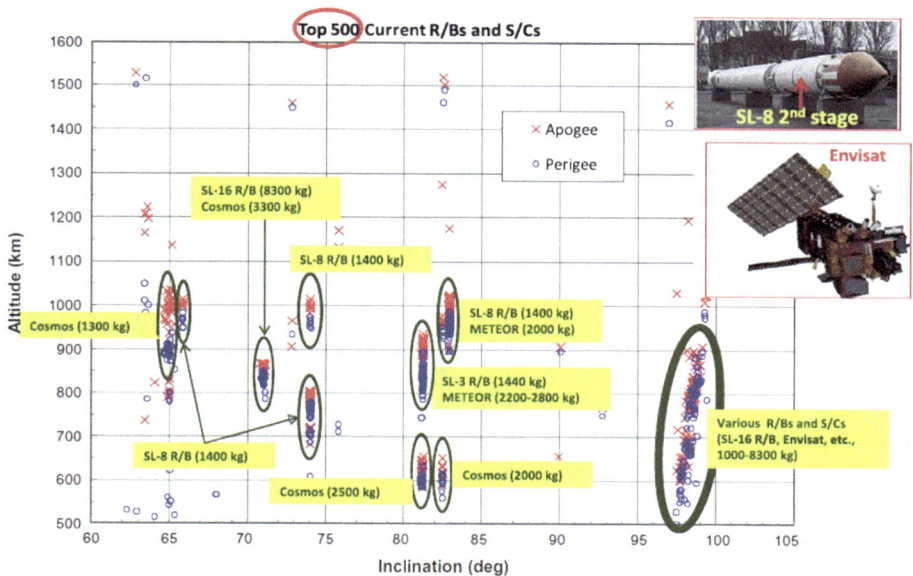

8.2 Sources of orbital debris tend to exist in families located at various altitudes and orbital inclinations. Source: J. C. Liou, *An active debris removal parametric study for LEO environment remediation*. Advances in Space Research 47 (2011) 1865–1876.

---

[2] Detailed information on orbital debris damage to the Shuttle fleet can be found in http://ntrs.nasa.gov/archive/nasa/casi.ntrs.nasa.gov/20110015922.pdf

on removing the larger pieces before they collide. Many of these larger objects are upper stage spent rocket casings, such as the Ukranian Zenit upper stage, which are tumbling uncooperatively in LEO in a variety of orbital inclinations. There are complex legal and jurisdictional issues associated with any effort in active debris removal (ADR), but regardless of the bureaucracy, the economics of such a clean-up task are strongly tied to rocket propulsion. Maneuvering a large, uncooperative object to a disposal orbit requires fuel, so the cost of the operation increases linearly with the amount of propellant required. Electric propulsion, with its frugal fuel consumption as compared with chemical rockets, is ideal for this application, and a high power solar-electric rocket such as VASIMR® would produce excellent results.

With these ideas in mind, in 2012, the Ad Astra Rocket Company developed its concept for an orbital sweeper, a 200 kW solar-electric VASIMR® space tug, capable of removing large pieces of orbital debris from LEO by controlled deorbit. The sweeper would be capable of deorbiting, in a single mission, up to 19 known large uncooperative objects in 19 different orbital planes. A company study examined the removal of a known family of existing 8-ton spent Zenit upper stages which are drifting in various high inclination orbits. This multipurpose space tug would be fitted with a solid rocket motor (SRM) tray, loaded with 20 SRM units (19 for the debris plus one reserved to deorbit the empty tray) and a detachable short-range "chemical robotic pod," or service module (SM), for specialized proximity operations near the target body. The VASIMR® tug lowers the orbital altitude of the targets and robotically plants a single SRM on each for a controlled chemical deorbit over the Pacific Ocean. When the disposal is complete, the tug climbs back up and executes an orbital plane change to rendezvous with another target. The mission involves a sequence of 19 altitude change maneuvers, in multiple orbital planes, optimized for minimum fuel use and minimum time for a given power and specific impulse (I_{sp}). The Ad Astra study found an absolute minimum time at an I_{sp} of 4500-5000 seconds, well suited for a VASIMR® propulsion system operating with low-cost argon propellant. Lower I_{sp} results in a large increase in the initial mass in low Earth orbit (IMLEO) and therefore increased cost.

THE *OCELOT*™ SOLAR-ELECTRIC POWER AND PROPULSION MODULE

Another important application that has been envisioned since the early days of the company is the re-boost of large orbiting stations in LEO, and its variant called atmospheric drag compensation. At just under 400 km above the Earth, the ISS orbits in a region where tiny amounts of the planet's atmosphere are still present. This results in a small amount of drag on the vehicle, much of which is exerted on its sizeable solar panels, truss structure and pressurized modules. The atmospheric drag is sufficient to alter the orbiting outpost's speed (about 27,000 kph) and this reduction in speed cumulatively results in a drop in altitude. This drag is greater during periods of increased solar activity, because the Sun heats and expands the upper atmosphere, increasing the local density where the ISS flies and thus accentuating the effect.

To manage this altitude loss and prevent an unrecoverable orbital decay condition, the station requires regular altitude boosts. Over the years, these have been provided by visiting spacecraft, such as the Russian Progress vehicle, the European Automated Transfer Vehicle (ATV) and, of course, the Space Shuttles while the fleet was still in service. With the Shuttles and the ATV retired, orbit re-boosts must now be performed by the Progress. In all cases though, the re-boost is performed with conventional chemical propulsion. At its present altitude, the estimated requirements for ISS are in the order of about 3500 kg of re-boost fuel per year. This fuel has to be transported from Earth, both at a cost and at the expense of useful payload mass. This re-boost service could be commercially provided with a high power solar-electric propulsion module that operates continuously. The much higher I_{sp} of the electric rocket would significantly reduce the chemical re-boost fuel up mass and the VASIMR® engine is particularly well suited for this application.

Ad Astra's *Ocelot*™ solar-electric power and propulsion module is designed to provide autonomous re-boost to a variety of LEO space platforms, such as the ISS. The 100 kW class VF-100™ VASIMR® engine module features independent solar power generation, with a dedicated high power solar array and standard docking capabilities, and can safely perform orbital re-boost and drag compensation for a tenth of the cost of conventional chemical propulsion. *Ocelot*™ is delivered to the station by a commercial transport vehicle. Once docked, *Ocelot*™ can trim its thrust vector to compensate for center of gravity offsets on the customer's station, insuring a linear, non-rotational maneuver. Its open "extension collar" architecture and standard docking interfaces at both ends enables full docking to other visiting spacecraft and unimpeded pressurized tunnel access through the module to the rest of the customer's station.

8.3 Ad Astra's *Ocelot*™ solar-electric power and propulsion module concept, capable of re-boosting space platforms such as the ISS.

BUILDING A CISLUNAR TRANSPORTATION SCAFFOLDING

As interest grows in revisiting – and establishing permanent settlements on – the Moon and points nearby, a robust and economical logistics transportation capability will be required. For example, a space station could be built at the first Lagrange point (L1)[3] between the Earth and the Moon as a strategic depot for supplies, fuel, equipment, food, water and so on, in support of human habitation on the Moon's surface or as a staging point to build and supply deep space ships. Therefore, the transportation of payloads beyond the Earth's active satellite constellation is also of great interest. In this realm, the VASIMR® engine facilitates a potentially lower cost transportation option in cislunar space, an advantage that stems directly from the reduction in fuel mass per kilogram of payload delivered.

As an example, one could ask: From a total initial mass in low Earth orbit (IMLEO) of fifty tons, how much could one deliver to the L1 Lagrange point? A short study of a conceptual 400 kW VASIMR® solar-electric space tug was conducted by Ad Astra in 2013, with operational scenarios at three different values of specific impulse. These missions were compared to two all-chemical alternatives, one at an I_{sp} of 450 seconds, characteristic of cryogenic LOX-Hydrogen, and the other at an I_{sp} of 350 seconds, characteristic of less energetic chemical fuels. Table 8.1 shows the resulting mass budgets for the various options and the time involved. Inspection of the data shows that the electric option results in two- to three-fold improvement in the payload delivered over the chemical option from the same IMLEO. The high I_{sp} option results in the highest payload but also the longest transit time. The chemical options assume one way flights only and hence no reusability of the upper stage. These result in a low payload, albeit with the fastest transits.

From this, an interesting transportation architecture could be envisioned for cislunar space, one that combines slow but high payload trans-lunar electric tugs with less efficient, but faster chemical transport for humans. These considerations bring into focus the importance that robotic, high-power solar-electric tugs with VASIMR® propulsion could have in economically delivering supplies to the Moon and strategic depots near our natural satellite. Such an approach should receive high priority, as it would enable the implementation of a robust transportation "scaffolding," supporting long-term human habitation beyond LEO.

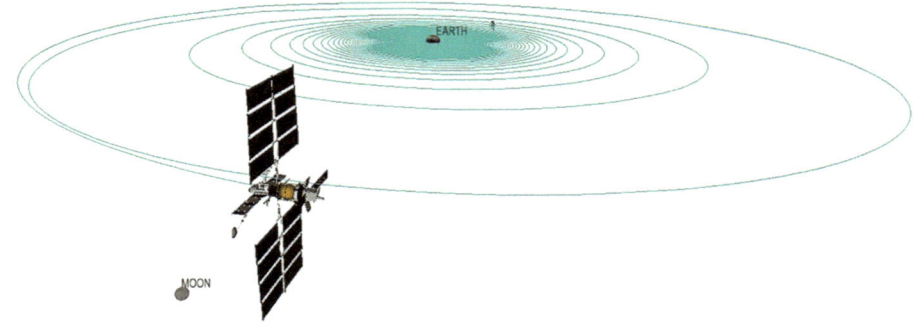

8.4 Lagrange Point L1.

---

[3] Lagrange points are locations where the pull of gravity from various nearby celestial bodies cancel each other out. In this way, an object placed there remains in equilibrium.

Table 8.1   Ad Astra's 400 kW Cislunar Solar Electric Space Tug Study.

Definitions for the table are shown below, all values in metric tons (t). Assumptions in the study are as follows: Mass of the fuel tank is 10 percent of the mass of the propellant. The spacecraft dry mass less tankage is assumed to be 12kg/kW.

Initial mass in LEO	**IMLEO**	Final mass after returning to LEO	**FMLEO**
Mass of propellant used from LEO to L1	**Prop (LEO-L1)**	Outward duration from LEO to L1 (days)	**LEO-L1**
Mass of payload for space station delivery	**Payload**	Return duration from L1 to LEO (days)	**L1-LEO**
Initial mass after delivery at L1	**IML1**	Mass expelled per second as thrust	**mdot**
Mass of propellant used to return from L1 to LEO	**Prop (L1-LEO)**	Change in velocity from LEO to L1	**DelV**

Isp	Mass Budget (t)						Time (days)		mdot	DelV [m/sec]
(sec)	IMLEO	Prop (LEO-L1)	Payload	IML1	Prop (L1-LEO)	FMLEO	LEO-L1	L1-LEO	[kg/sec]	LEO-L1
5000	50	6.3	37.5	5.6	0.7	4.8	363	41	0.00020	6,556
2500	50	12.0	30.3	6.5	1.6	4.8	173	22	0.00080	6,652
1500	50	18.8	21.1	8.2	3.1	4.8	98	16	0.00222	6,811
450	50	29	15	6	chemical one way only		4	N/A	N/A	3800
350	50	33	10	7	chemical one way only		4	N/A	N/A	3800

IN-SPACE RESOURCES

The third area of commercial interest is the new field of space resource utilization, includ-ing the potential salvage of spacecraft material and the mining of asteroids and near-Earth bodies, which may harbor valuable natural resources including water. Mining of these relatively small Sun orbiting bodies will likely require their commercial repositioning to more stable and readily accessible locations, such as the vicinity of the Moon. NASA's Asteroid Redirect Mission (ARM) attempted to study this problem, albeit with a signifi-cantly underpowered spacecraft. Ad Astra conducted similar mission studies with a much higher power propulsion system, featuring VASIMR® engines in the 200 to 400 kW range. To enable a useful comparison, the Ad Astra study considered the 1300 ton, 2008HU4 near-Earth asteroid (NEA) proposed in the recent Keck Institute for Space Studies (KISS) asteroid retrieval study [1]. However, instead of a 40 kW Hall thruster, operating with xenon gas, the company utilized a more powerful (100–400 kW) VASIMR® propulsion system, operating with either of two significantly cheaper propellants: argon or krypton (see table 8.2).

The results show the importance of faster delivery of the target asteroid to minimize the escalating costs caused by the time value of money in a commercial operation. By confin-ing the SEP technology to the Hall thruster, the KISS study produces a significantly under-powered, 10-year mission, costing $2.6 billion in 2012 dollars, not including the cost of the xenon propellant or a consideration for the time value of money. An increase in SEP power from 40 kW to 100–400 kW results in lower cost and faster delivery. A 255 kW VASIMR® produces a four-fold future cost improvement and a five-fold reduction in the mission time [2].

A variant of the asteroid retrieval scenario is a mission solely dedicated to planetary pro-tection. The February 15, 2013 asteroid explosion over the Russian city of Chelyabinsk brought this heavenly threat into sharp focus, as thousands witnessed the airburst of an 18-meter diameter rock that had made its fiery entry through the Earth's atmosphere. At 25

Table 8.2 Ad Astra Asteroid Retrieval Study Results.

Name	KISS	VF-150	VF-200	VF-300	VF-400	VF-255
Power	40 kW	150 kW	200 kW	300 kW	400 kW	255 kW
Propellant Type	Xenon	Argon	Argon	Argon	Argon	Argon
Specific Impulse (Isp)	3000 sec.	3000 sec.	3000 sec.	4000 sec.	5000 sec.	3400 sec.
Efficiency	60%	61%	61%	68%	73%	65%
Propulsion System Mass (KISS pg. 26)	1.0 t	1.0 t	1.0 t	1.0 t	1.0 t	1.0 t
Power System Mass (KISS pg. 26)	1.1 t	3.3 t	4.4 t	6.5 t	8.5 t	5.5 t
Asteroid Capture System Mass	0.2 t	0.2 t	0.2 t	0.2 t	0.2 t	0.2 t
Tank Mass (4% of propellant, as in KISS)	0.5 t	0.6 t	0.6 t	0.4 t	0.4 t	0.6 t
Spacecraft Dry Mass (KISS pg. 26)	5.3 t	7.7 t	8.7 t	10.6 t	12.6 t	9.8 t
Mass in LEO (KISS pg. 26)	18.9 t	22.2 t	23.6 t	21.6 t	21.6 t	23.0 t
Delta V (KISS pg. 29)	6.6 km/s	6.6 km/s	6.6 km/s	6.6 km/s	6.6 km/s	6.6 km/s
Propellant Used	4.03 t	4.72 t	5.03 t	3.54 t	2.87 t	4.37 t
Time	2.3 yrs	0.7 yrs	0.6 yrs	0.4 yrs	0.4 yrs	0.5 yrs
Heliocentric Delta V (KISS pg. 29)	2.8 km/s	2.8 km/s	2.8 km/s	2.8 km/s	2.8 km/s	2.8 km/s
Propellant Used	1.41 t	1.65 t	1.76 t	1.29 t	1.08 t	1.56 t
Time (w/50% coasting)	1.8 yrs	0.5 yrs	0.4 yrs	0.3 yrs	0.3 yrs	0.4 yrs
NEA Stay Time (days)	90	90	91	92	93	94
Total Propellant to NEA	5.4 t	6.1 t	6.5 t	4.7 t	3.9 t	6.1 t
Total Time to NEA	4.3 yrs	1.5 yrs	1.2 yrs	1.0 yrs	0.9 yrs	1.1 yrs
NEA Mass (KISS)	1300.0 t	1300.0 t	1300.0 t	1300.0 t	1300.0 t	1300.0 t
Heliocentric Delta V (KISS pg. 29)	0.17 km/s	0.17 km/s	0.17 km/s	0.17 km/s	0.17 km/s	0.17 km/s
Propellant Used	7.7 t	7.7 t	7.7 t	5.8 t	4.6 t	6.8 t
Time (w/25% coasting)	6.0 yrs	1.6 yrs	1.2 yrs	0.9 yrs	0.8 yrs	1.0 yrs
Chemical Propellant (KISS pg. 29)	0.4 t	0.4 t	0.4 t	0.4 t	0.4 t	0.4 t
Total EP Propellant Used	13.1 t	14.1 t	14.5 t	10.6 t	8.6 t	12.7 t
Total Mission Time	**10.1 yrs**	**3.1 yrs**	**2.4 yrs**	**2.0 yrs**	**1.8 yrs**	**2.0 yrs**
Power System (KISS pg. 40)	$251M	$417M	$492M	$643M	$794M	$575M
Thruster (KISS pg.40)	$224M	$288M	$308M	$342M	£371M	£328M
Spacecraft (KISS pg. 40)	$1,243M	$1,472M	$1,568M	$1,753M	$1,933M	$1,671M
Payload Cost (KISS pg. 40)	$93M	$93M	$93M	$93M	$93M	$93M
Contractor Fee (10% without PL)	$115M	$138M	$148M	$166M	$184M	$158M
Spacecraft w/fee (KISS pg. 40)	$1,357M	$1,610M	$1,716M	$1,919M	£2,117M	$1,829M
NASA (15% of SC, KISS pg. 41)	$204M	$242M	$257M	$288M	$318M	$274M
Phase A (KISS) (KISS pg. 41)	$68M	$81M	$86M	$96M	$106M	$91M
Mission Ops (KISS pg. 41)	$116M	$51M	$45M	$41M	$40M	$42M
Launch Vehicle: Atlas 551 (KISS) or Delta IV	$288M	$300M	$300M	$300M	$300M	$300M
Reserve (30%)	$610M	$685M	$721M	$793M	$864M	$761M
Total (minus propellant)	**$2,643M**	**$2,969M**	**$3,125M**	**£3,438M**	**$3,744M**	**$3,297M**
Propellant (Xe: $1000/kg, Ar: $5/kg)	£13M	$0.07M	$0.07M	$0.05M	$0.04M	$0.06M
Total + Propellant with 20% Time Cost of Money	**$19,980M**	**$5,495M**	**$5,074M**	**$5,091M**	**$5,364M**	**$4,948M**

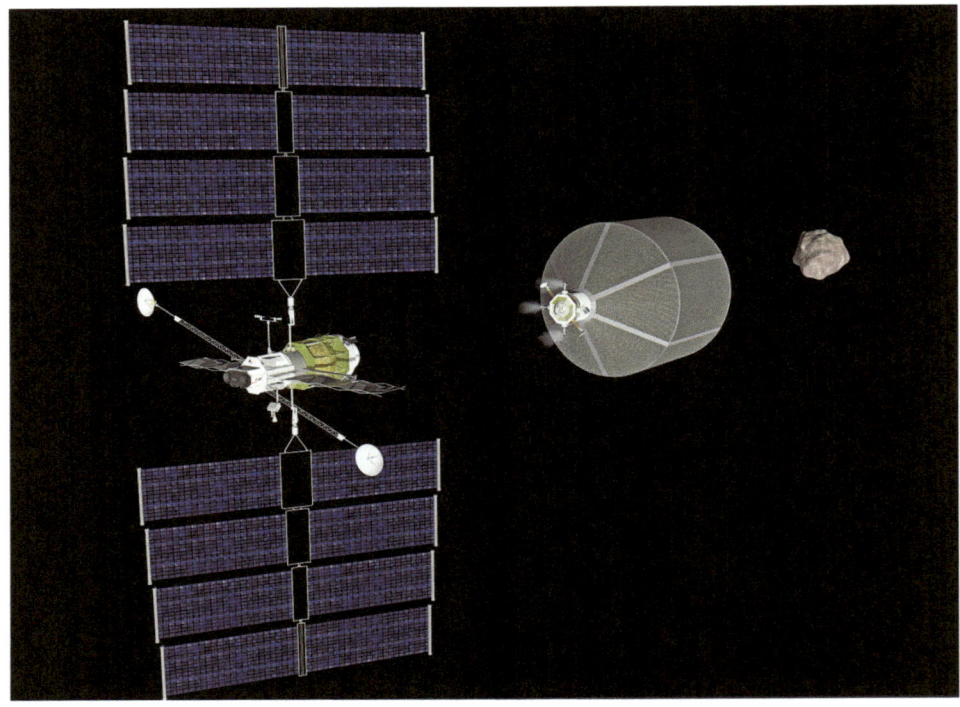

8.5 Concept of a 200 kW VASIMR® engine adapted to the KISS study NEA retrieval module.

kilometers above the ground, the asteroid exploded with the force of a nuclear blast that was fortunately high enough to cause only minor injuries and damage to buildings. Nonetheless, the violence and unpredictability of the event has raised worldwide awareness of the severity of the asteroid threat to our home planet. Indeed, other such impacts and close calls have shaped our world and have had major effects on our biosphere in the past, including the formation of our Moon 4.5 billion years ago, the extinction of the dinosaurs 65 million years ago and the leveling of 2,150 square kilometers of forest in Tunguska, Siberia, in 1908. We are able to detect and track the very large asteroids and can confidently say that the Earth is not presently under an extinction level threat from above. We are also confident that our atmosphere is sufficiently thick to incinerate the small asteroids before they reach the ground. We are, however, in the blind when it comes to the medium size asteroids, such as was the case in Chelyabinsk. Dr. Edward Lu, a former NASA astronaut who has been studying asteroids for more than a decade, points out that there are about one million asteroids in the Earth's vicinity "with the potential to destroy a major metropolitan area."

Several Planetary Defense (PD) techniques have been proposed, from violent direct kinetic impactors and nuclear blasts, to the gentler – and longer-term – nudging by strategically positioned electric thrusters and gravity tractors, which require a warning of the threat well in advance of the potential impact. Ad Astra Rocket Company has evaluated the applicability of its 400 kW solar-electric propulsion (SEP) "space tug" on the latter approach with a conceptual asteroid deflector spacecraft, which the company has named *Viento*™.

Viento™ is a 400 kW solar-electric spacecraft equipped with two dual-core, VF-200™ -class engines operating at 200 kW each. The robotic craft features two hinged engine nacelles that can be configured both in additive *translation mode* to provide high speed translation capability to the target asteroid, and in *deflection mode* with the engine modules oriented in opposing tandem. In this latter mode, the plasma exhaust of one of the engines is used to gently "blow" on the asteroid to impart momentum and alter its trajectory, while the other engine, with proper gimballing and auxiliary spacecraft attitude control, maintains the spacecraft in a stable position, hovering adjacent to the asteroid without actually landing.

The company has conducted a mission simulation of *Viento*™ by digitally creating an imaginary, 150-meter diameter and 7-million-ton asteroid, called *Khan*, and placing it in a similar trajectory as that of the known asteroid, 99942-Apophis[4]. Unlike the orbit of Apophis, which will result in a near miss with Earth on April 13, 2029, the orbit of *Khan* has been slightly modified to result in a direct impact[5]. *Khan* is in a nearly circular orbit

---

[4] 99942-Apophis is a 270 m diameter boulder weighing approximately 40 million tons and was discovered on June 19, 2004.

[5] In a direct impact (vs. a "keyhole" impact), an asteroid is on an orbital track that will result in a certain collision with Earth on a specific date. A keyhole impact also results in a collision with Earth, but only following a prior near-miss, during which the asteroid passes through a "keyhole," a small region near Earth where the gravitational force of the planet causes a change in the asteroid's orbit, setting up a subsequent direct impact. Avoiding a keyhole is less demanding than a direct impact and could potentially be accomplished using a Gravity Tractor. Avoiding a direct impact requires a robust deflection capability.

with a period of 323 days. Each year, Earth crosses *Khan*'s orbit on April 13 as Khan heads inbound toward perihelion. *Khan* is large enough to pose a major threat to our planet and in our imaginary scenario, if not deflected, *Khan* will impact Earth with an energy release of 131 megatons, causing a major regional disaster.

Ad Astra's *Viento*™ carries out the deflection campaign in four phases:

1) *Departure* on August 13, 2019 from Earth-Moon L1 (EML1) and a 305-day propulsive translation to a rendezvous with *Khan* on June 13, 2020.
2) A five-year active *deflection* period, ending on June 13, 2025, where the spacecraft is configured to hover adjacent to the asteroid while pushing on it with the other engine.
3) A four-year passive *loiter* period at *Khan*, ending on March 19, 2029, while *Viento*™ awaits an optimal return opportunity.
4) A 40-day *return* maneuver, which brings *Viento*™ back to its point of origin at the EML1.

The VASIMR® deflection capability is determined by the power level, the deflection time and the size and mass of the asteroid. At the 400 kW level used in the simulation, the deflection of a 150 m asteroid is readily facilitated within the allotted time from a direct impact at the center of the Earth out to several Earth radii. Larger asteroids, up to ~200 meters in diameter, may also be deflected just enough to avoid a collision.

8.6 Concept of a >300 kW VASIMR®-SEP space tug with attached payload.

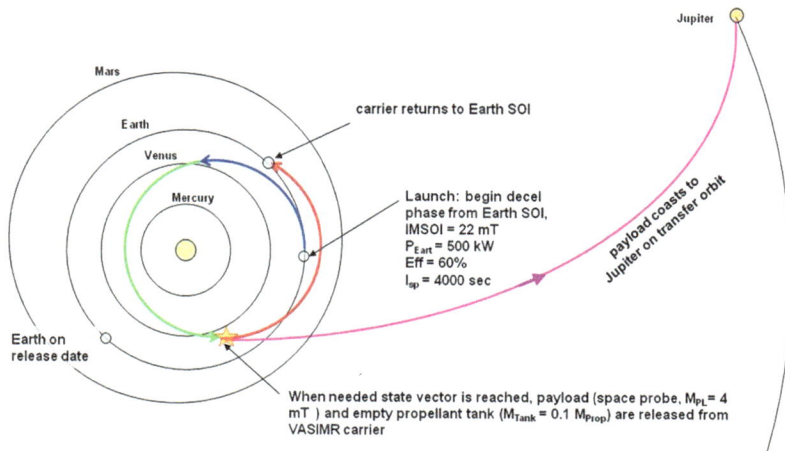

8.7 Ad Astra's VASIMR® powered Deep Space Catapult solar electric space tug would be capable of delivering large scientific payloads to points in deep space very quickly. Once sufficient velocity is achieved, the tug releases the payload, which coasts to its destination.

FAST DELIVERIES TO THE DEPTHS OF THE SOLAR SYSTEM

Another important application of the high power VASIMR® engine is Ad Astra's Planetary Catapult[6]. The robotic solar electric space tug is designed to deliver large (>2 Mt) scientific payloads to points in deep space very fast, by means of a high power (> 300 kW), solar electric power system.

Instead of relying on gravity assists, which may not be available for the desired flight opportunity, the mission architecture enables the VASIMR® engine to deliver sufficient propulsive impulse to the space tug and its payload in an accelerating trajectory arc of high solar illumination in the inner solar system. The proximity to the Sun reduces the required size of the solar arrays or the need for solar concentrators, making the overall spacecraft significantly lighter. Once sufficient velocity is achieved in the high illumination arc, the tug releases the payload, which coasts to its destination in the outer solar system. The space tug returns to Earth orbit under its own power and is recovered for multiple uses. In this way, the VASIMR® propulsion system could provide fast and affordable primary propulsion for a growing market of deep space planetary missions carrying exploratory robots. In deep space science missions, a large portion of the mission cost is associated with the significant period that an active science team must wait for the probe to reach its destination. Moreover, the reusability of the tug also implies that its launch cost can be amortized over several missions, while payload capability, in a multi-mission program, could be significantly enhanced.

---

[6] An animation of this mission, showing the various phases of flight and the evolution of trajectory and performance parameters is available at http://www.adastrarocket.com/aarc/jupiter-catapult

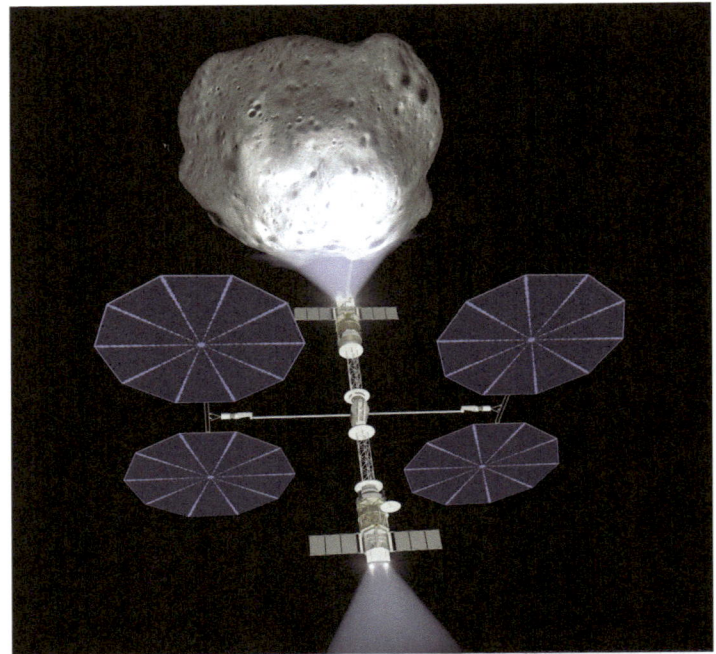

8.8 Ad Astra's *Viento*™ spacecraft is configured in two modes: additive propulsive mode (left) for fast transit to the asteroid, and opposing tandem mode (right) for deflection operations. The plasma plume from one of the engines exerts a small but continuous force on the asteroid, independently of its rotation, while the other engine maintains the deflector at a stable distance. The 400 kW solar electric propulsion spacecraft is equipped with two, VF-200-class VASIMR® engines.

REFERENCES

1. J. Brophy, et al. Asteroid Retrieval Feasibility Study, Keck Institute for Space Studies, Pasadena, CA, Apr 2 2012. http://kiss.caltech.edu/study/asteroid/asteroid_final_report.pdf.
2. A. V. Ilin, VASIMR® Solar Powered Missions for NEA Retrieval and NEA Deflection, IEPC-2013-336, 33rd International Electric Propulsion Conference, Washington, DC, Oct 6-10, 2013.

9

Mission Threats and Potential Solutions

Space is hard. It is a dangerous and unforgiving place and all astronauts know that space missions are laden with risk. Nonetheless, humans in general still long to go to space and astronauts in particular, while keen to avoid unnecessary danger, never shy away from acceptable risks. But there is a difference between taking risks for risk's sake and taking risks because, in accomplishing a higher purpose, there is no other choice. You could climb a mountain by slowly clawing with your bare hands up a steep and perilous wall; one false move on your way to the top and you will fall to the abyss and die. Or you could take a helicopter to get to the top quickly and safely because your end goal is not the climb, but to reach the summit to accomplish a task.

These considerations illustrate the two extremes in the motivational continuum for space exploration and they frame much of the discussion about going to Mars and points beyond. A half century ago, during a tumultuous era of confrontation that brought us to the brink of nuclear war, the United States embarked on a journey to the Moon. Its motivation combined fear and nationalistic pride, mixed in the correct proportions to win an open race for technological supremacy in the heavens. The massive national effort produced the desired result: humanity's most amazing technological achievement; and then, it was over.

Fifty years later, the sociopolitical chemistry of our planet today is very different from that of the 1960s. Confrontation has given way to globalization, the internet, the democratization of science and technology and the recognition that our planet's resources are finite and access to them is disrupting a fragile ecosystem that supports all life on Earth. It is clear now that our survival depends on collaboration. We must look to space exploration as a race with ourselves to ensure the survival of our species. The time for nationalistic stunts and chest pounding is over. It is we humans of all nations that must ensure a steady and sustainable outward motion of our civilization into space. Working together, we can indeed, in the words of the late President John F. Kennedy, "organize and measure the best of our energies and skills." A lasting lesson from Apollo: In the haste to reach the summit, let us not forget why we chose to climb. This time, we would be well advised to invest in the right technologies now and not have to wait another fifty years to move forward after someone plants a flag on Mars.

© Springer International Publishing Switzerland 2017
F. Chang Díaz, E. Seedhouse, *To Mars and Beyond, Fast!*, Springer Praxis Books,
DOI 10.1007/978-3-319-22918-8_9

9.0. Scott J. Kelly, right, in 2011, with the Russian astronauts Oleg Skripochka, left, and Aleksandr Kaleri, after six months aboard the International Space Station. There won't be any recovery teams waiting for crews on the surface of Mars. Credit: Bill Ingalls/NASA

A human mission to Mars is doable today. It was also doable 20 years ago, and it will be in 2035, the latest estimated date. The collective enthusiasm to land humans on the Red Planet, on a three-year mission using conventional propulsion technology, has not diminished. Presenters at conferences routinely explain how astronauts can survive cancer-inducing radiation and how over-the-horizon technology can protect crews from bone deconditioning and muscle atrophy. Myriad simulation missions and analogues, such as HI SEAS, MARS500, FMARS and MDRS have been conducted to keep alive the dream of humans on Mars.

Yet most diehard Mars fans acknowledge that a human mission using conventional propulsion is a risky proposition that lies on the ragged edge of today's capabilities. The two fundamental technological enablers for such a mission, power and propulsion, have changed little since the Moon landings. Under these conditions, the colonization of the Red Planet would be unnecessarily fragile and, as the Apollo program taught us, the sustainability of a technologically fragile exploration program is questionable in the long run. Fortunately, the technologies required to bring sustainability and robustness to such a program are not as far away as they appear.

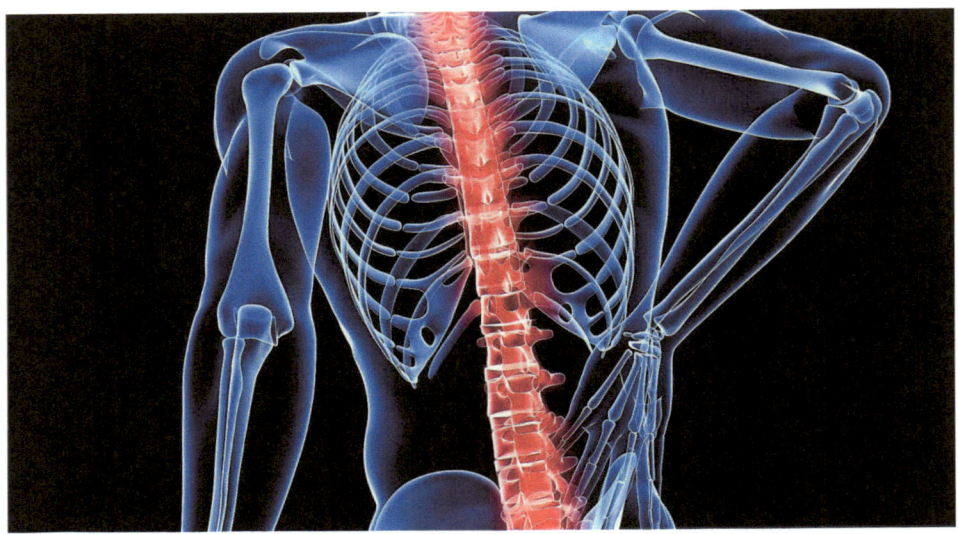

9.1. Osteoporosis presents a serious threat to Long Duration Exploration Missions - LDEMs. Credit NASA

THE RISKS OF VENTURING FURTHER AFIELD

Deep space exploration poses many risks that add extraordinary threats to an already challenging mission, as they affect the general health and safety of astronauts. While they will not be eliminated entirely, these risks can be greatly reduced by adopting, without further delay, the space equivalent of the "Nautilus Paradigm." Maintaining a healthy crew, able to perform in top physical and mental condition, is paramount to giving them a "fighting chance" of successfully dealing with the myriad known and unknown risk factors associated with the journey. We have not ventured far in the fifty years since Apollo, but we have learned a great deal about living and working in space, and all of these lessons point to the urgent need to develop a faster and more energetically robust transportation system.

The most serious risks to those venturing beyond Earth orbit (BEO) appear to lie in the subatomic realm, and for Mars-bound astronauts that means radiation. Galactic Cosmic Rays (GCRs) are the nuclei of heavy atoms bulleting through space at extraordinarily high velocity. The particles can penetrate most materials, including human tissue. They have the potential to give deep space astronauts a dose of radiation serious enough to shred genetic material (see table 9.1) and cause cancer and possibly death. At the same time, solar bursts of high energy charged particles, known as Solar Particle Events (SPEs) but more commonly referred to as "solar flares," also pose a threat. SPEs tend to be fairly random, although they are less active during the more peaceful periods in our star's 11-year cycle. The Mars rover *Curiosity*, whose mission was planned to coincide with one of these quieter periods, recorded five SPEs.

9.2. Deep space radiation is strong enough to seriously damage DNA. Credit NASA

During a trip to the Red Planet using conventional propulsion, the radiation dose a crewmember might receive is 0.66 Sieverts[1] (a unit that measures the physiological effect of radiation). That is a lot of radiation, considering that an accumulated dose of one Sievert is the lifetime limit for an astronaut. Radiation is a mission threat that varies with an astronaut's medical history, gender and age. Radiation absorption during a trip to the Red Planet could raise the risk of cancer by up to five percent, a figure that exceeds NASA's

---

[1] A report published in the May 2013 edition of *Science* estimated that crews embarking on a 360-day return trip to Mars would be exposed to 662 ± 108 millisieverts (mSv) of radiation exposure. The study indicated as much as 95 percent of the radiation would come from Galactic Cosmic Rays (GCRs), which are hard to block without using a substantial amount of shielding.

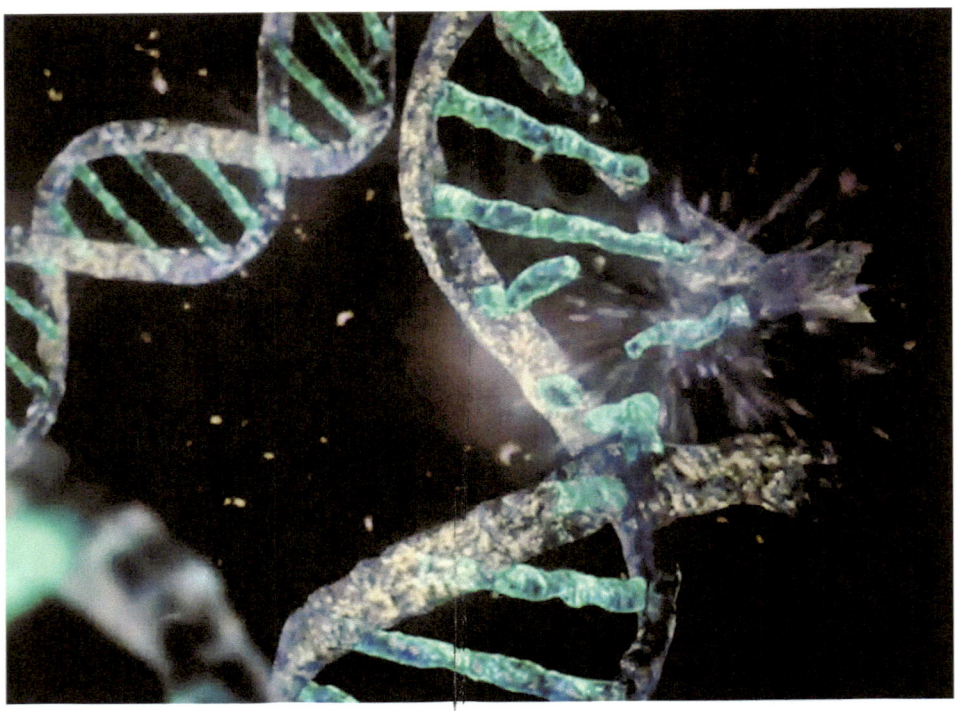

9.3. Solar flares are unpredictable and deep space crews will have little or no warning of SPEs. Credit NASA

Table 9.1 **Symptoms of radiation sickness.**[A]

EARLY	**Mild Exposure (1-2Gy)**	**Moderate Exposure (2-6Gy)**	**Severe Exposure (6-9Gy)**	**Very Severe Exposure (10Gy or higher)**
Nausea/vomiting	Within 6 hours	Within 2 hours	Within 1 hour	Within 10 minutes
Diarrhea	N/A	Within 8 hours	Within 3 hours	Within 1 hour
Headache	N/A	Within 24 hours	Within 4 hours	Within 2 hours
Fever	N/A	Within 3 hours	Within 1 hour	Within 1 hour
LATER				
Dizziness/disorientation	N/A	N/A	Within 1 week	Immediate
Weakness, fatigue	Within 4 weeks	Within 1 – 4 weeks	Within 1 week	Immediate
Hair loss, bloody vomit and stools, infections, poor wound healing, low blood pressure	N/A	Within 1 – 4 weeks	Within 1week	Immediate

[A]Adapted from Radiation exposure and contamination. The Merck Manuals: The Merck Manual for Healthcare Professionals limits. Additionally, female astronauts have a 20 percent lower threshold for radiation exposure than men, due to an increased risk from breast, ovarian and uterine cancers. Of course, agencies could relax the cancer threshold, as there will always be plenty of astronauts queuing up to take the risk. The problem is that radiation is a capricious hazard. Its long-term effects are still not clearly understood and may affect the immune system, cause problems with short-term memory, and increase the risk of heart disease. Some recent concerns have surfaced over potential Alzheimer-like symptoms [1]. For multi-year missions, these medical issues may occur during the mission itself, when adequate treatment may not be readily available.

Cosmic rays may also cause mutations. When astronauts launch, they will likely be carrying microbes. As cosmic rays zip through the crew, the rays may mutate these microbes, with the result that the bugs may reproduce faster and become stronger. The cosmic rays will also exert a significant stress on the immune systems of the astronauts, with the result that crewmembers might become more susceptible to illness. With limited medical care on board, a spate of illnesses is a formidable mission threat. More than 100 research studies have analyzed the effects of space flight on astronauts and pathogens and the results point to a potentially immuno-compromised crew during the course of the mission.

Recently, in 2014, the results of two NASA studies – Validation of Procedures for Monitoring Crewmember Immune Function (Integrated Immune) and Clinical Nutrition Assessment of ISS Astronauts, SMO-016E (Clinical Nutrition Assessment) – published in the *Journal of Interferon and Cytokine Research* [2], have provided evidence that long duration missions may temporarily affect the function of the immune system. NASA's Integrated Immune study revealed that cell function in those astronauts on board the ISS is depressed and when this happens, the immune system does not respond to threats as efficiently as when it is functioning normally. One effect of a malfunctioning immune system is *viral shedding*, which means dormant viruses in the body may be stimulated, leading to a spate of allergies and rashes. This is not something that astronauts would want to deal with on a multi-year mission. More evidence of immune system malfunction was revealed when NASA scientists studied the blood of 28 ISS crewmembers prior to, during, and post-flight. In this study, the researchers were particularly interested in examining cytokine function. Cytokines are proteins that regulate immunity by recruiting immune cells to fight an infection. It is a process more commonly known as inflammation. In long duration missions, NASA found evidence [3] that blood cytokines are altered in such a way that the effectiveness of the inflammatory response is degraded. This means that astronauts en route to Mars could find themselves at a higher risk of infection. The current state of research indicates that there is no stabilization of the immune system; in other words, it is impossible to say if the problem would get better, or worse.

On the positive side, radiation exposure can be mitigated during flight by utilizing material shielding rich in hydrogen (such as water and a number of plastics). It has been calculated that about 100 centimeters of water or polyethylene shielding might be sufficient to stop most of this radiation [4]. In the VASIMR® ship, the bulk of the proton radiation from the Sun could be effectively shielded by the liquid hydrogen making up the propellant. This propellant could be strategically located in cryogenic storage tanks surrounding the crew module. In addition, the liquid hydrogen could enable the installation of a nested toroidal superconducting magnetic shield, which will further increase the radiation protection. Shielding astronauts on the surface of Mars could be accomplished by a

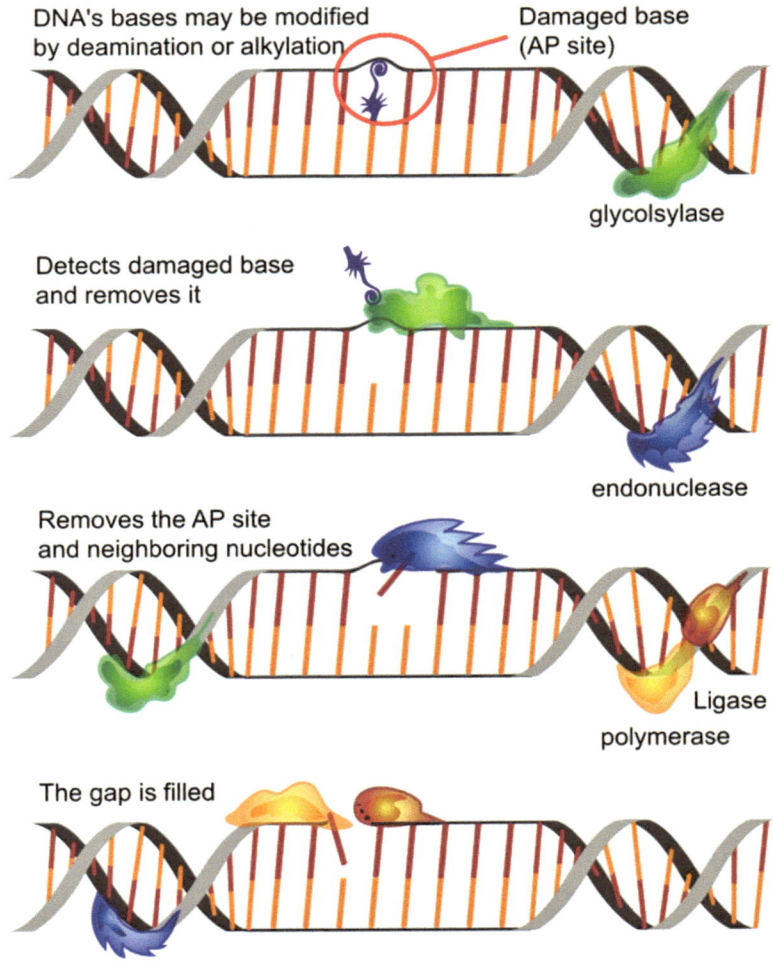

9.4. Houston: we have *another* problem! DNA under attack. Our genetic material – DNA – comprises four letters (nucleotides); A, T, C and G. Combined, these nucleotides form genetic code. Each time a cell grows and divides, every gene must be copied to the next generation of cells. The problem is that this DNA replication process can be seriously threatened by deep space radiation, which can inflict several thousand genetic damage events on each cell every day! The body has an intricate mechanism capable of repairing and restoring the original genetic sequence but it can only do so much, because there is a limit beyond which the body's recognition machinery breaks down. When it does break down, mutations are the likely result. Credit NASA

variety of methods, including water layers around the crew habitat and taking advantage of the terrain itself. Rock formations near the landing site could provide very effective barriers to high energy particles and mission planners will probably study these options well in advance of the flight.

The increased cancer risk is a subject of great debate among mission planners. Many astronauts are willing to accept the increased risk as an occupational hazard. Nonetheless, arguably few would do so if they had the choice of a faster mission. The medical community is also working on a protocol of antioxidants and non-steroidal anti-inflammatory drugs. It is possible that natural agents such as beta-glucan, curcumin, lycopene, and ferulic acid[2] could cut the risk. Possible, but not likely in view of the degree of radiation exposure that the astronauts would face during a three-year mission. Astronauts could also take concoctions of radiation protectants, such as WR-2721, a drug that protects astronauts from gamma radiation. The problem is that WR-2721, while effective, has a number of adverse side effects, including toxicity and vision impairment.

NASA is taking the cancer problem very seriously. It has research efforts underway that are investigating individual sensitivity to cancer. They are researching biomarkers, generating sophisticated model projections of radiation risk, modeling shielding for various mission durations, working on dosimetry countermeasures, testing new pharmaceutical and nutritional intervention strategies, and developing personalized cancer screening biomarkers. The agency is also investigating new radiation shelters, mission care, and, notably, new and faster propulsion systems, because NASA acknowledges that one of the best ways of reducing cancer risk is to reduce transit times.

Whether radiation induced or not, space brings about other physiological changes that cause concern, including reduced visual acuity and cataracts. Post-flight examinations performed on about 300 astronauts since 1989 showed that 29 percent of Shuttle crewmembers (who flew two-week missions) and 60 percent of ISS astronauts (who spent four to six months in orbit) experienced a degradation of visual acuity[3]. Visual impairment has only recently been reported on orbit. Initially, researchers thought the problem was due to an increase in pressure on the optic nerve. This pressure, combined with the effect of elevated carbon dioxide levels inside spacecraft (about 20 times normal Earth levels of .23 mmHg), was thought to be the likely main cause, but other researchers have shown that different mechanisms may be implicated. Post-flight examination of astronauts has revealed optic disc edema, sphere flattening, choroidal folds and reports of increased intracranial pressure. Some astronauts have experienced transient changes following their flight, while others have suffered lasting symptoms.

On the subject of cataracts, a NASA study [6] has revealed that long duration astronauts are more likely to acquire them. The study, which investigated 48 cataract cases in astronauts, not only discovered a marked increase in cataracts for those who had been exposed

---

[2]Beta-glucan is found in the cell walls of yeast and is a potent protector against radiation damage in tissues, especially bone marrow. Curcumin, which is an anti-carcinogenic extract of turmeric, protects cells against every form of radiation damage. Lycopene is a carotenoid and a powerful antioxidant that stimulates production of antioxidants and protects against radiation. Ferulic acid increases the amount of protective antioxidants in the body and improves the recovery of bone marrow stem cells [5].

[3]*U.S. National Academy of Sciences*

to high levels of (space) radiation, but also noted that astronauts who had been exposed to higher levels of radiation suffered cataracts at a younger age than those crewmembers who were exposed to lower radiation levels. The study also suggested that long duration crewmembers were at higher risk for cataracts than short duration astronauts. Much research is ongoing to alleviate this potential mission threat.

The musculoskeletal problems associated with space flight were discovered early in the human space program. Some progress has been made to reduce this mission threat, but the weakening of the bones and the loss of muscle tissue associated with long exposure to weightlessness continue to be of great concern.

> "Bone is a living tissue, and must be 'stressed' to maintain strength. If bones are immobile for long periods, as occurs in space but also in bedridden patients, the individual will lose a substantial amount of muscle and bone mass, which may have serious repercussions."
>
> *Professor Rizzoli, chairman of the IOF Committee*
> *of Scientific Advisors, speaking at the IOF World Congress on*
> *Osteoporosis, Toronto, Canada, 2-6 June, 2006*
> *2006 Scientific Committee*

Bedridden patients and long duration space-farers suffer a common medical condition: progressive bone loss. Terrestrial-bound patients lose bone density due to inactivity, while those on orbit lose it from the absence of weight on the skeleton. This bone loss is continuous and progressive, and, along with radiation, is one of the most significant mission threats. There have been hundreds of studies that have investigated the skeletal changes and loss of body calcium experienced by astronauts. High urinary calcium concentrations were reported in Skylab astronauts after only a few days on orbit. In several of the crew, those urinary calcium concentrations remained higher than normal for the duration of the mission. The body needs that calcium to build bone and if the body is instead getting rid of it, the bones weaken. These concerns can be exacerbated by the fact that the decalcification may not be completely reversible upon returning the body to a gravity field. The progress that has been made in reducing this mission threat through the development of skeletal loading countermeasures has been significant, but the condition remains a going concern.

Much of the ongoing research into this issue is devoted to trying to limit the damage. A special sclerostin antibody is being considered as a potential treatment for stemming the tide of bone loss. Sclerostin is implicated in bone loss and if there is some way of stopping its activity, the rate of bone loss might be reduced. Dietary strategies, including meals rich in fish oils and Omega-3 fatty acids may also be useful. Other, mechanical countermeasures are also being proposed, such as the Gravity Loading Countermeasure Skinsuit (GLCS). This garment works – thanks to its bidirectional elastic-weave material – on the premise that bone responds to loading. The Skinsuit provides mechanical pressurization for the body, so bone loss can be reduced, or at least that is the theory. As with all the mission threats discussed here, we will have to wait a little longer to know for sure.

Closely related to decalcification and bone loss are kidney stones, a condition that is exacerbated by the natural dehydration of the human body in weightlessness. Kidney

9.5. Without the presence of Earth's gravity, astronauts lose 1 to 2 percent of their bone mass per month on the ISS, especially in the load-bearing spine and lower limb bones. This rate of loss is up to 15 times faster than observed loss rates in serious osteoporosis patients and is no good if you happen to be a prospective Mars astronaut! But good news is on the way: the European Space Agency (ESA) is developing the Gravity Loading Countermeasure Skinsuit (GLCS), which may prove a valuable countermeasure against bone loss by mimicking the effects of gravity. It has been designed by a consortium consisting of the Massachusetts Institute of Technology, Kings College London, University College London, and Wyle Laboratories GmbH. Credit. ESA.

stones are an occupational hazard for astronauts because it is nigh on impossible to keep hydrated while working in space. Compounding the hydration challenge are the side effects of bone demineralization, which causes an elevation of salt levels in the blood. This in turn results in higher concentrations of salts in the urine, which in turn increases the risk for developing kidney stones. Making matters worse is the difficulty in trying to locate the stone, because interplanetary spaceships will not be equipped with the resources needed to do this. One option is for patients to drink water in the hope that the stones pass naturally, but this method is not always successful.

"Once renal stones start to move they can be excruciatingly painful. You'd have an incapacitated crew member and potentially have to abort the mission."

Peggy Whitson, NASA astronaut and biochemist

The skeleton will not be the only physiological system that takes a beating as astronauts make their way to and from Mars. Consideration must also be given to muscle loss, paying particular attention to those muscles that surround the spinal column and help you stand upright. Without these muscles, gravity would collapse you into a fetal ball and leave you curled up on the deck. Sculpted by the force of gravity, these muscle groups are in a state of permanent exercise, constantly being loaded and unloaded. In space it is a different story, because up there those particular muscles are among the fastest-wasting groups in the body.

As with other conditions, researchers have studied the muscles of astronauts who have spent six months on board the ISS. The problem, again, is the lack of load or pressure on the muscles. In short, if there is no weight, the muscle fibers are unloaded. This is a problem because it is this loading that maintains muscle size, so the lack of load leads to atrophy. Studies have analyzed muscle fibers from calf biopsies of crewmembers before and after their missions to see what happens. Single muscle strands were isolated and assessed for their ability to generate force. The results show the muscle fibers lost a third of their capacity to do so.

An important problem upon arrival on the surface of Mars after a prolonged journey in weightlessness will be orthostatic intolerance. This problem arises from a sudden drop in blood pressure, or *orthostatic hypotension*, and it occurs in nearly all astronauts after long-term missions. In a gravity environment, your autonomic nervous system kicks in when you stand, adjusting your cardiovascular system to maintain the correct blood pressure. But when someone with orthostatic hypotension stands up, their autonomic nervous system does not respond as it should and blood pressure drops, depriving the brain of blood and oxygen. The person feels dizzy and may faint. Long duration astronauts landing on Earth experience the same effects, which is why they are carted off on stretchers. While astronauts don't suffer orthostatic intolerance until they return to Earth, once on the ground, the condition is challenging. Astronauts feel weak and have to sit down in the shower, for example. As a precautionary measure, space agencies do not allow their long duration astronauts to drive for three weeks following return from orbit. That time may be reduced on Mars because the surface gravity there is 38 percent that of Earth, although having to bear the weight of a fully loaded EVA suit on Mars may offset that advantage.

Beyond radiation, bone loss and the other mission threats considered here, defiance, detachment, disagreement and sheer loneliness may turn out to be more threatening conditions. There can be unpredictable emotions, with unpredictable responses, in any small group, but during a multi-year mission to Mars, such feelings could be life threatening. While it is true that astronauts undergo rigorous personality testing, many of those tests only provide an approximate measurement of how personalities change in a confined and isolated space. In the end, these issues also merit important consideration and indeed they are being seriously addressed now. Clearly, embarking on a round trip to Mars using chemical propulsion is a challenging proposition that behooves us to consider a faster alternative.

LIFE SUPPORT AND CREW SAFETY

There are, of course, myriad other mission threats associated with the interplanetary ship itself and all of its systems. Of particular importance is the life support system; the ability to cleanse the environment and reduce disposable consumables. In the Space Shuttle, the oxygen and hydrogen supplies were consumed to produce water, electricity and breathable air for the crew. Waste water was not recycled but dumped overboard and the carbon dioxide produced by the crew was chemically removed with a finite amount of lithium hydroxide filters, the supply of which often determined the ultimate length of the mission.

Technology has advanced considerably since the early Shuttle flights and, for the ISS, CO_2 removal can be accomplished much more sustainably. In addition, waste water can also be recycled and some of it can be electrolyzed to produce breathing oxygen as well as hydrogen, with the latter typically vented overboard for safety. Other processes could also be invoked to further improve on a truly sustainable atmosphere by collecting the waste CO_2 and combining it with the hydrogen from electrolysis to produce water and methane (the Sabatier reaction). All of these systems, however, require electrical power, a resource that presently comes from the Sun. The power system of a Mars vehicle is, therefore, extremely critical and its architecture must provide ample redundancy and robustness to accommodate unforeseen failures. Unlike the fairly constant solar illumination that a spacecraft enjoys while in cislunar space, the amount of solar energy available en route to Mars drops quickly (as the square of the distance) as the ship travels away from the Sun. Sunlight at Mars is about half as bright as it is on Earth. If solar power is used, solar arrays must be sized to fully support the needs of the ship and its crew at Mars orbit. Nuclear power, on the other hand, has no such constraint.

Human space missions must have abort capability and provide reasonable means of safely returning the crew back to Earth in the face of critical system failures. All crewed missions flown to date, including those of Apollo, have never truly left Earth orbit. The spacecraft have remained "bound" to the home planet, even while in the local gravity well of the Moon. In all these missions, a survivable return after a major contingency, while extremely challenging, can be contemplated and, indeed, executed – as with Apollo 13. The Mars ship, on the other hand, after its trans-Mars propulsive burn is complete, is bound to the Sun or to Mars. In a contingency, the ship must first reach Mars and, under the right conditions, including adequate fuel and consumables remaining, utilize the planet's gravity to bend the trajectory back towards the inner solar system and a potential rendezvous with Earth. Such abort scenarios need to be contemplated and evaluated for reasonable survivability and in many cases, protecting for these contingencies will have strong implications on the architecture and scope of the mission.

From the foregoing discussion, it is evident that power and propulsion are the two most critical elements in a space ship and, as we shall discuss in Chapter 10, the "Nautilus Paradigm" of high power nuclear-electric propulsion greatly widens the option space for a survivable mission.

REFERENCES

1. Cherry, J.D., Liu, B., Frost, J.L., Lemere, C.A., Williams, J.P., Olschowka, J.A., O'Banion, M.K. *Galactic Cosmic Radiation Leads to Cognitive Impairment and Increased Aβ Plaque Accumulation in a Mouse Model of Alzheimer's Disease.* PLOS one. December 31, 2012.
2. Crucian, B.E., Zwart, S.R., Satish, M., Uchakin, P., Quiriarte, H.D., Pierson, D., Sams, C.F., Smith, S.M. *Plasma Cytokine Concentrations Indicate That* In Vivo *Hormonal Regulation of Immunity Is Altered During Long-Duration Spaceflight* J. Intereron Cytokine Res. 2014 Oct 1; 34 (10): 778-786.
3. https://humanresearchroadmap.nasa.gov/evidence/reports/Immune_2015-05.pdf http://phys.org/news/2014-08-reveals-immune-dazed-spaceflight.html
4. Cucinotta, F.A., Myung-Hee, Y.K., Chappell, L.J., Huff, J.L. *How Safe is Safe Enough? Radiation Risk for a Human Mission to Mars.* PLoS One. 2013; 8 (10).
5. Fang YZ, Yang S, Wu G. *Free radicals, antioxidants, and nutrition.* Nutrition. 2002 Oct; 18(10):872-9.
6. Nelson, E.S., Mulugeta, L., Myers, J.G. *Microgravity-Induced Fluid Shift and Ophthalmic Changes.* Life 2014, 4, 621-665

10

The VASIMR® Nuclear-Electric Mission Architecture

As discussed briefly in Chapter 4, the VASIMR® team instigated a number of conceptual Nuclear-Electric Propulsion (NEP) Mars mission studies, starting in the early 1990s and using trajectory codes available at the time that could be adapted for constant power throttling (CPT) conditions. The 1995 NASA report "*Rapid Mars Transits with Exhaust Modulated Plasma Propulsion* [1]," was the first such study to examine the advantages of high power NEP coupled with VASIMR® technology to enable a new class of fast missions, with potential for further improvement through advances in the technology.

The VASIMR® team continued this work during the transition from the ASPL to the Ad Astra Rocket Company, exploring a number of additional mission scenarios that further enhanced the earlier work during the first decade of the 21st Century. For example, the team considered utilizing the Earth-Moon Lagrange points as potential strategic locations for the construction of the Mars vehicle and for safer assembly of the nuclear reactors. They also continued to examine the abort capability of the propulsion system in response to failures in two of its critical elements; the propellant storage and the reactors. The team does not consider these studies to be an exhaustive assessment of mission operations, but instead views them as motivational illustrations of important areas that will continue to require closer attention.

FIRST VASIMR® OPTIMAL TRAJECTORIES UNDER VARIABLE I$_{SP}$

The first in-depth study of optimized trajectories with VASIMR® propulsion was carried out by a team at NASA's Johnson Space Center (JSC) under the leadership of Dr. Chang Díaz. The group included Mr. Michael Hsu, at the time a midshipman from the US Naval Academy at Annapolis who was detailed to the Astronaut Office for a summer, as well as Ellen Braden and Ivan Johnson, two JSC orbital dynamicists. Ellen and Ivan were

© Springer International Publishing Switzerland 2017
F. Chang Díaz, E. Seedhouse, *To Mars and Beyond, Fast!*, Springer Praxis Books,
DOI 10.1007/978-3-319-22918-8_10

implementing a number of modifications to an existing NASA trajectory code to incorporate variable I$_{sp}$, also known as "exhaust modulation," the optimal approach employed by the VASIMR® engine. Mr. Hsu was assigned to work with Dr. Chang Díaz and his team at the ASPL and was given the task of using the modified code to examine the conditions required to accomplish fast transits to Mars with VASIMR® NEP.

In 1995, the concept of exhaust modulation was not new. It had been known, at least theoretically, since the early 1950s and had been thoroughly described by Dr. Ernst Stuhlinger [2], one of the early rocket pioneers in Wernher von Braun's team working in the US Rocket Program in Huntsville, Alabama. For a fixed transit time, the technique yields maximum payload capability through an optimum propulsive schedule of thrust and I$_{sp}$, one which must also ultimately consider the local strength of the gravitational well in which the vehicle is traveling. One major advantage of this is the flexibility in the level of propulsive "muscle" available to the ship at any given moment in order to extract the best performance; not only during the cruise phase of the journey where gravity fields are likely to be low, but also upon reaching its destination where the tug of gravity is stronger.

However, aside from its more rudimentary form in chemical rockets, requiring multiple engines at different fixed I$_{sp}$ and commonly known as "staging," the rocket pioneers knew that the practical application of the full extent of exhaust modulation in a single engine was not possible with the technology available in the 1950s. The technique was therefore left as a potential area of work for future generations, once technological advances could make it possible. Such advances had become available to the VASIMR® team by the 1980s and soon after they began their exploration of variable I$_{sp}$ rockets, they came across the pioneering work of Dr. Stuhlinger and his team.

The 1995 study analyzed basic Mars mission scenarios within the familiar "split-sprint" architecture envisioned by numerous previous studies. In such a scenario, a one-way, slow, automated cargo ship leaves first and places a habitat, fuel, and supplies at strategic points in Mars orbit and on the planet's surface. Following this robotic, high-payload mission, a smaller, low-payload fast ship carries the crew to Mars. The fast ship picks up its final fuel load for Mars at a staging rendezvous point at the edge of the Earth's sphere of influence (ESOI) and heads for Mars under a CPT schedule at maximum power to Mars orbit insertion. Rendezvous with the Mars lander and supplies is necessary to proceed to a landing.

In this early and highly simplified work, no attempt was made to consider the gravitational fields of the Earth and Mars. In essence, the ship travels in heliocentric space under the sole gravity of the Sun between the orbits of Earth and Mars, departing and arriving with zero velocity at either end of the journey. The ship is propelled by a 10 MW VASIMR® NEP with an alpha value of 6 kg/kW. At two percent payload mass fraction, the human mission sacrifices payload for speed and achieves a 101-day outbound trip, a 30-day stay on the surface, and a 104-day return. This "speedboat" option was the shortest Mars mission possible for the parameters assumed in the study. In contrast, the robotic cargo ship arrives at Mars in 180 days with a 67 percent payload mass fraction. Longer missions would result in higher mass fractions.

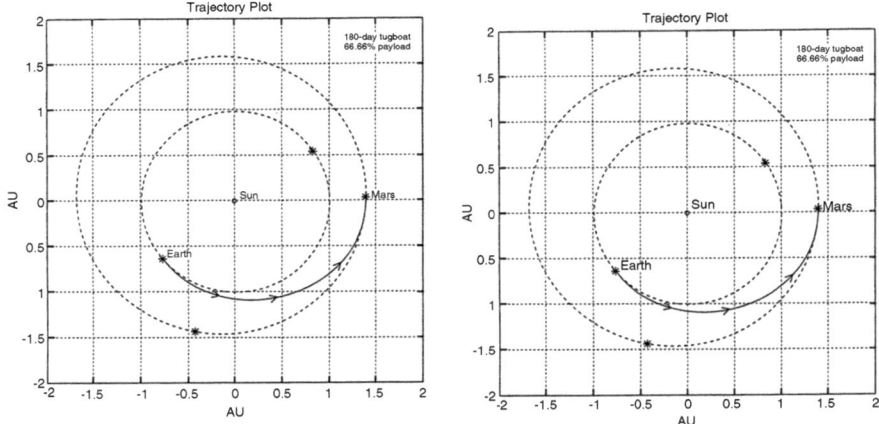

10.1 Split-sprint architecture includes a slow robotic cargo ship (left) followed by a fast piloted "speed boat." Total mission duration includes a 30-day surface stay with nearly symmetrical inbound and outbound legs of approximately 100 days. Shorter human transits are allowed by the propulsion system, but the favorable alignment of the planets for the transit necessitates an extended 2-year stay on the surface.

The relative motion of Earth and Mars in this scenario negates symmetrical outbound/inbound heliocentric transit flights of less than 104 days in the same Martian year. However, symmetrical outbound/inbound one-way flights of 90 days or less are possible if the crew can remain on the surface for a Martian year (two Earth years). Such missions would be unrealistic for the first mission, but would certainly be possible once a robust permanent outpost exists on the planet.

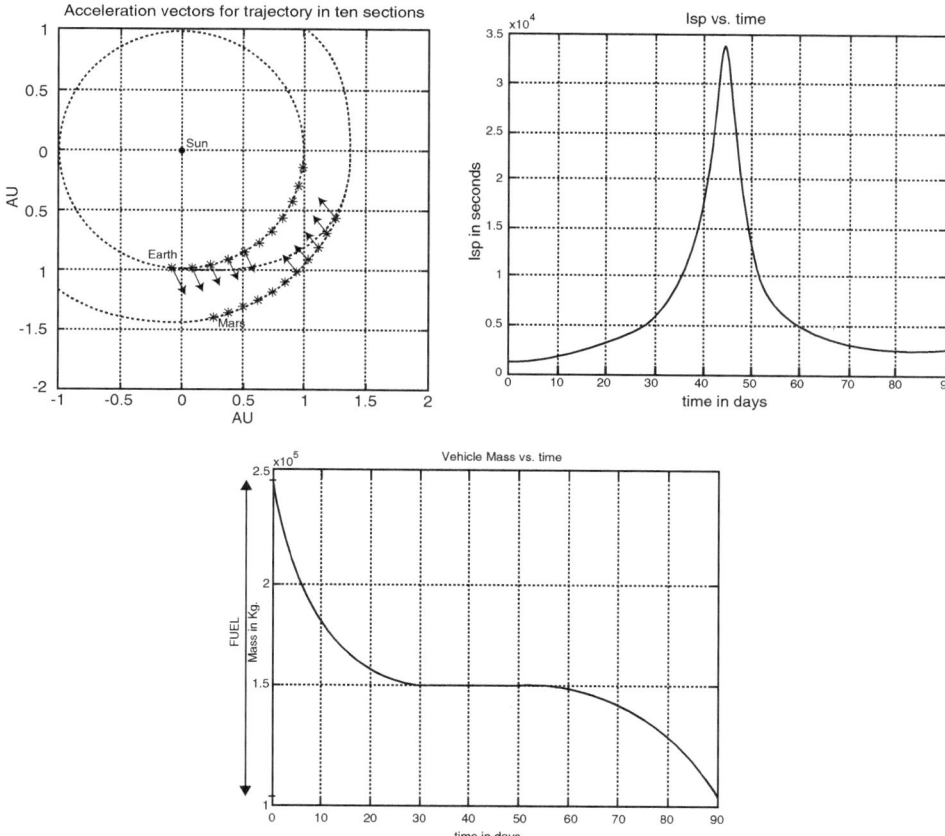

10.2 The thrust acceleration for the fast transits (left) is nearly radial and decreasing toward the trajectory midpoint, reversing direction, and increasing as Mars approaches at Mars arrival. I_{sp} (right) and thrust are exchanged continuously at constant (maximum) power. The I_{sp} profile is nearly symmetrical, with minor differences due to lower Sun gravity and reduced total mass at Mars arrival. Fuel expenditure (bottom) is high during the initial and final high-thrust phases of the mission but quite small in the middle, suggesting that a truncated "coast" could be implemented without the need to throttle to an unreasonably high I_{sp}.

EARLY ABORT SCENARIOS

Abort capabilities, essential for human missions, were also explored in the 1995 study, albeit only considering very simplified assumptions: for example, what the options would be if the crew habitat were to fail or the crew became incapacitated; and what could be done if one or more of the modules comprising the propulsion cluster were to fail. Under several of these conditions, the system would have a limited capability to return directly to Earth with an increased travel time, or alternatively could abort to Mars if a suitable safe haven could be secured there.

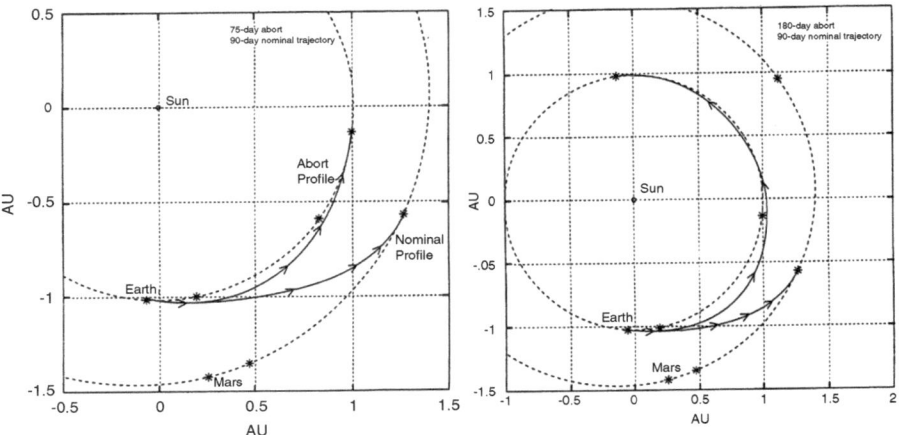

10.3. A powered abort with a full-up propulsion system 15-20 days into the mission enables a 75-day return to Earth. A 180-day abort (right) could result from a propellant system failure, reducing the fuel available and hence increasing the return time.

The abort scenarios considered in the 1995 study assumed that a full-up propulsion system was still available. Other abort scenarios were considered in later studies by the VASIMR® team and will be discussed shortly. However, in the 1995 study, the abort was arbitrarily declared at 15 days into a 90-day mission. This yielded two possible return-to-Earth options, at 75 and 180 days. The 180-day option represented the potential loss of propellant, most likely due to a plumbing failure or a leak from one or more of the fuel tanks. In general, because of the capability to throttle to a high I_{sp}, the efficiency of fuel usage increases with the abort duration.

Two other VASIMR® NEP trajectory studies were completed between 1998 and 2011, at the ASPL and later at the Ad Astra Rocket Company. Some of these results were presented at the University of Maryland as a compilation entitled "VASIMR® Human Missions to Mars"[1]. The tools employed in these two exploratory investigations included the NASA-JSC Hybrid Optimization Technique (HOT) utilized in the 1995 study and "Copernicus," a generalized spacecraft trajectory design and optimization system developed for NASA in 2002 by Dr. César Ocampo of the University of Texas at Austin. "Copernicus" remains the most advanced tool available to date.

---

[1] Space, Propulsion & Energy Sciences International Forum, March 15-17, 2011, University of Maryland.

The simulations utilizing the HOT trajectory code featured a 188 mT ship departing from low Earth orbit (LEO) on May 6, 2018, including a 61 mT Mars Lander comprising a 31 mT Habitat, a 14 mT Aeroshell and a 16 mT chemical descent system. The propulsion package was a 12 MW VASIMR®-NEP system with a more optimistic alpha value (than the 1995 study) of 4 kg/kW. These studies produced a one-way transit time to Mars of 3-4 months, a two-fold improvement over the conventional chemical mission being considered at the time.

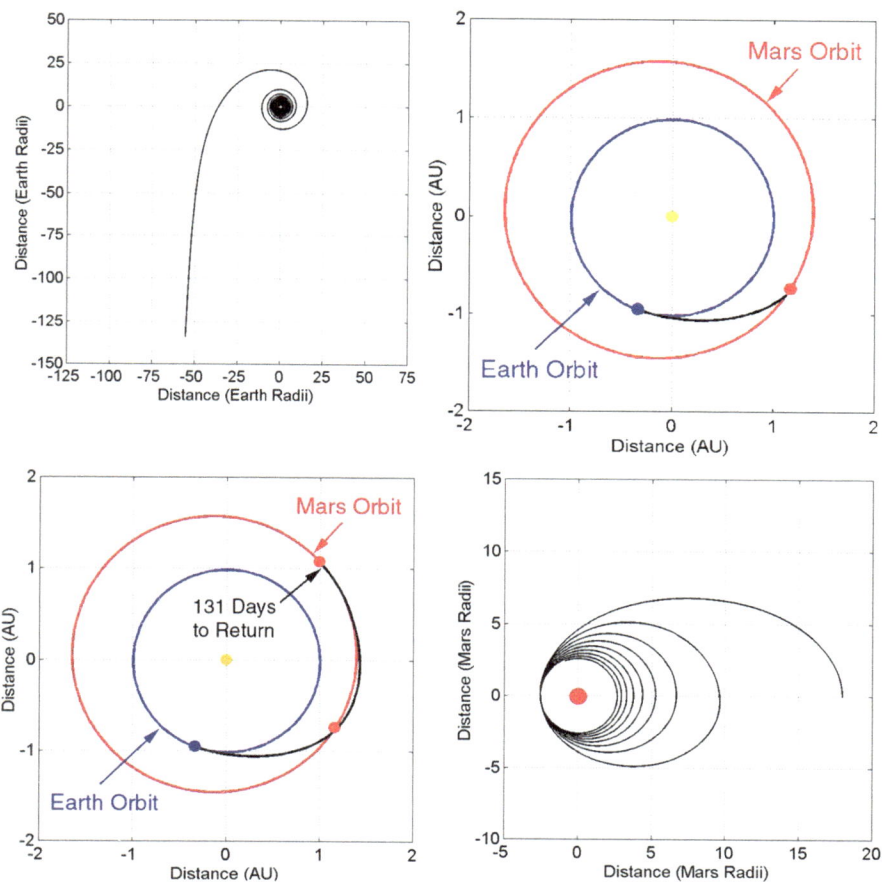

10.4. Human piloted mission high thrust spiral (upper left), heliocentric transfer (upper right), CTV arrival at Mars and subsequent capture 131 days later (lower left) and 7-day spiral maneuver into low Mars orbit (lower right).

The mission included a spiral trajectory from LEO and a heliocentric transfer to Mars. At Mars arrival, the mother ship and lander would separate, allowing the former to descend to the surface and the latter to execute a fuel-saving grazing flyby of the planet and continue past Mars, prior to a second rendezvous and Mars capture approximately four months later. Upon its second Mars arrival, the mother ship would gradually spiral down to a low Mars orbit (LMO) and dock with the cargo ship parked in orbit, allowing access to the return propellant and habitat. The cargo engine and reactor package could then be released and either be left in Mars orbit for future use or sent on a disposable trajectory to the Sun. Once so configured for the return flight, the mother ship would await the return of the crew from the surface.

Several assumptions were made in the design of this architecture, including the overall propulsion system efficiency of 60 percent, the arrival velocity at Mars of 6.8 km/sec and the execution of a direct chemical descent to the surface, with an identical descent maneuver to that envisioned by the NASA Design Reference Architecture (DRA) available at that time. The mission also assumed the "split-sprint" NASA architecture, in which a robotic cargo ship has prepositioned a number of assets on orbit and on the surface of Mars prior to the launch of the piloted mission. In this way, the surface infrastructure would be built up and pre-tested prior to crew arrival, while the return fuel, habitat and some non-perishable provisions would be pre-positioned in orbit by the Mars cargo ship.

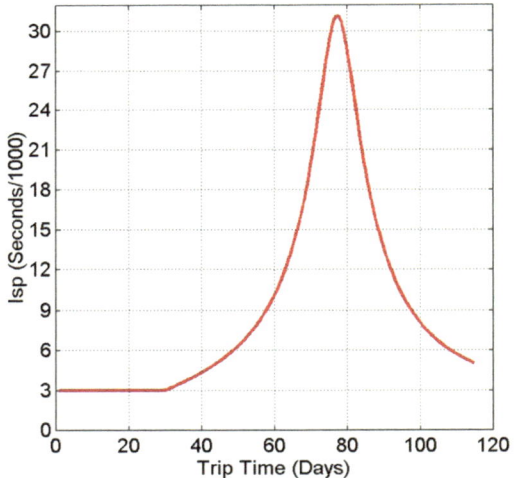

10.5. I_{sp} variation as a function of time. A constant I_{sp} (3000 sec), 30-day spiral, from Earth follows with upward (accelerating) and downward (decelerating) ramps. The theoretical I_{sp} peak of 30,000 is probably not required and a cut off can be established at 10,000–15,000 sec for practical implementation.

The lander separation from the mother ship at Mars arrival, and its direct entry, were designed to provide the propulsion module with ample opportunity to achieve orbital insertion at Mars, thereby saving a considerable amount of fuel and time. However, the automatic docking of the mother ship with the cargo vehicle carrying the return fuel and habitat could fail. If that were the case, the crew could opt to utilize the fully functional, albeit less powerful, VASIMR® propulsion module on the cargo ship. In such a contingency scenario, the return trip would take longer.

Crew exposure to radiation is always a concern and reducing exposure to Van Allen belt radiation still requires a short journey time, even with the significant radiation protection provided by the VASIMR® hydrogen propellant considered in this study. The scenario considered a 30-day Earth spiral, followed by an 85-day heliocentric transfer. The crew vehicle would operate at a constant I_{sp} of 3000 seconds until the vehicle departed the ESOI. Mars arrival would occur at a relative velocity of 6.8 m/sec and the Lander would execute a direct descent to the surface while the "mother" ship continued past Mars, as discussed earlier. The I_{sp} schedule used in the piloted mission retained a similar shape to that of in the 1995 study, with the addition of a 30-day constant I_{sp} and thrust climb to leave the Earth gravity well.

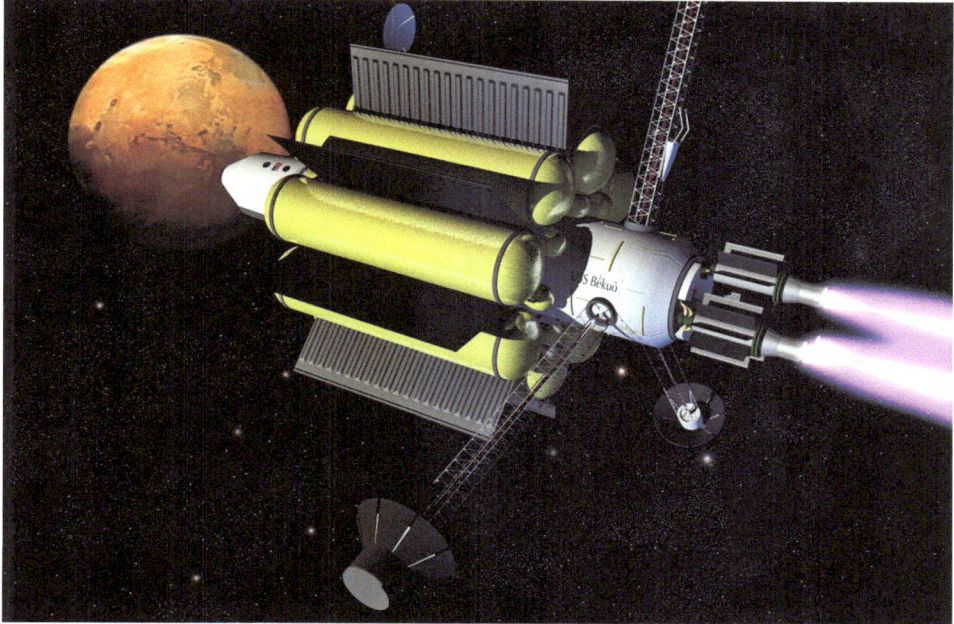

10.6. Earliest rendition of the VASIMR® NEP Crew Transfer Vehicle concept. Liquid hydrogen tanks surround the crew habitat for radiation protection throughout the mission. Three nuclear reactors at the end of deployable booms generate 12 MW of electrical power. Large radiators are required to shed the substantial excess heat produced in the power conversion process.

Ascent from Mars for the return flight would be accomplished chemically, with an ascent capsule that could also be used for direct descent to the Earth's surface. The return mission would consist of a four-day Mars spiral followed by an 85-day heliocentric transfer. At mission completion, the propulsion module could either be stored in a stable high Earth or Moon orbit for future use, or abandoned on a disposal trajectory to the Sun.

Three abort scenarios were postulated in this architecture for a return to Earth, following: 1) a propellant system failure in which one third of the remaining propellant was lost on Day 39; 2) a non-propulsion related failure on Day 44; and 3) an abort deep into the heliocentric trajectory, which would result in a very long return journey and merit strong consideration of an emergency landing on Mars. For this last contingency, it would be unlikely that a crew would choose to abort to Earth on Day 80 of the heliocentric trajectory. However, a number of malfunctions could also occur, with the remaining systems on the Martian surface or on the vehicle itself, which would preclude a landing. In this case, the 430-day return would be the only option available. The disruptive innovation of the VASIMR® NEP technology is evident when one considers that, even in these extreme contingency scenarios, the return flight is still shorter than the two-year planetary cycle.

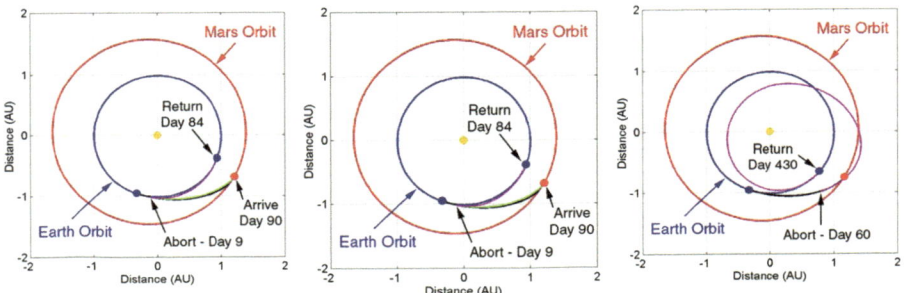

10.7. Abort to Earth from a loss of one third of the remaining propellant on Day 39 (left). Abort to Earth on Day 44 with a full-up propulsion system (center) and Abort deep into the heliocentric trajectory (right) resulting in a long return time.

FURTHER MODEL IMPROVEMENTS: COPERNICUS

As discussed in Chapter 5, after an extensive period of development and testing, the Copernicus generalized spacecraft trajectory design and optimization system developed by Dr. César Ocampo of the University of Texas at Austin became the "industry standard." This software is available to NASA centers and affiliates. It is a flexible tool, able to model different missions and gravitational fields. It can model multiple spacecraft operating under both constant and variable I_{sp}. Ad Astra Rocket Company uses Copernicus to explore the full range of VASIMR® NEP and Solar-Electric Power (SEP) applications.

In 2013, the VASIMR® team published a short study for fast and operationally robust human missions to Mars, based on high power NEP. The system considered an advanced, high temperature, gas cooled, nuclear magneto hydrodynamic (MHD) power plant, as proposed by Litchford and Harada in 2011 [3], combined with high power VASIMR® plasma propulsion. The general architecture of such a system featured a multi-megawatt closed cycle MHD nuclear plant, using non-equilibrium He/Xe frozen inert plasma (FIP) working fluid. The fission reactor mass estimate was based on the NERVA design for a 350 MW nuclear reactor with a mass of 1785 kg, increased to 3000 kg to account for containment and shielding margin.

For the radiators, the relevant design parameter is the mass per unit area of radiating surface. Articulating single-sided space radiators, in use today, typically operate at about 300 K and range from 6 to 10 kg/m². For the short study, the team considered two-sided, lightweight advanced thermal radiators, operating at 500 K, which can achieve a reduction by a factor of as much as two over the lower temperature, single-sided design. The higher temperature takes full advantage of the Stefan-Boltzmann law, which states that the radiated power per unit area is directly proportional to the fourth power of the radiator temperature. High temperature, two-sided radiators are relatively insensitive to the space environment near or beyond Earth's orbit. The upper limit for the heat rejection temperature is restricted by materials limitations and thermodynamic efficiency for the power conversion. Additional weight reduction can be achieved using carbon-carbon materials with very high thermal conductivity currently being developed. All of these advances could lead to future radiator specific mass values approaching 1 kg/m².

The VASIMR® portion of the system architecture extrapolates from the latest experimental results obtained on Ad Astra's 200 kW VX-200™ prototype. This fully superconducting system has demonstrated a thruster efficiency of 72 percent at specific impulse of 5000 sec, with argon propellant, at power levels up to 200 kW. Early experiments and modeling with krypton indicate that a multi-propellant VASIMR® system affords a potential envelope expansion to the higher thrust, lower specific impulse regimes without sacrificing efficiency, such that high efficiency operation could be possible over a range of specific impulse from 2000 sec to 5000 sec. With the exclusion of the RF power system and engine radiators, the mass for all other VASIMR® subsystems, for power levels between 200 kW and 2 MW, is estimated from VX-200™ and VF-200™ designs to be less than 500 kg. For higher power, the mass scales proportionally with input power P, due to clustering of the thrusters. The VASIMR® system mass is based on plasma model estimates and experimental data from the laboratory VX-200™ experiment, as well as projected masses for the 200 kW VF-200™ flight system design.

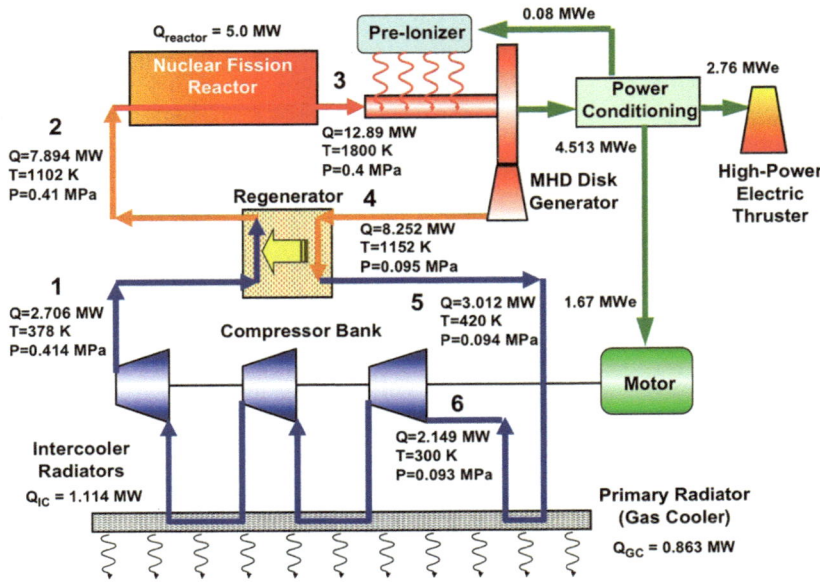

10.8. General architecture of an VASIMR®-NEP system with advanced MHD power conversion. Credit Litchford and Harada, 2011 [3]

NEP missions to Mars based on this power and propulsion configuration can outperform Nuclear-Thermal Rocket (NTR) missions for total specific mass values as high as 15 kg/kW. In addition, VASIMR®-NEP technology offers robust adaption to problem scenarios en route such as partial power or propellant failures, which can be addressed, albeit with an increased flight time, by an in-flight reconfiguration of the VASIMR® engine's specific impulse schedule.

The NEP mass model, shown in Figure 10.9, suggests that an advanced multi-megawatt nuclear VASIMR®-MHD system could approach α values of ~2 kg/kW for powers above 10 MWe. This framework for an integrated power/propulsion system has immense relevance to a fast human Mars mission architecture, even for α values of more than 10 kg/kW, and a preliminary evaluation of this concept was conducted in the 2013 study by a team from the Ad Astra Rocket Company, NASA and Nagaoka University in Japan. Particular attention was devoted to operational issues, such as major systems failure modes and mission abort scenarios, driven by power and propellant system failures en route.

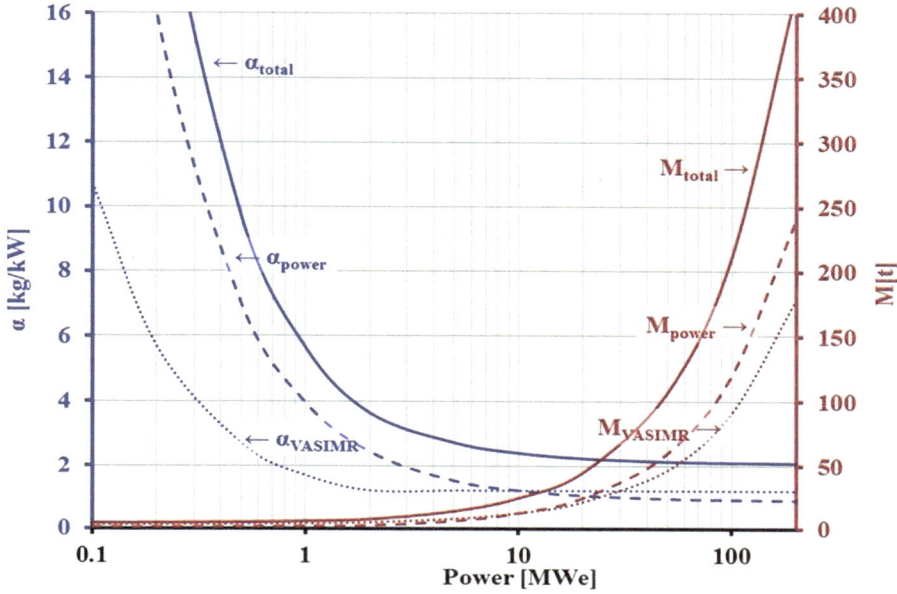

10.9. Specific and total mass for the nuclear VASIMR®-MHD system.

In a representative NEP mission scenario, the full vehicle departs from LEO (407 km) with an initial mass of 356.4 tons, including a payload (PL) mass of 62 tons. Upon arrival at Mars, the ship enters a one-sol orbit (250 km x 33,793 km) that enables a landing to be attempted. The NEP trajectory was optimized using the Copernicus interplanetary trajectory software. Figure 10.10 shows the resulting trajectory for a power level of 30 MWe and a total α of 2 kg/kW. The NEP mission results in a total in-flight time (out and back) of 149 days, which is 226 days shorter than the DRA-5.0. The shorter in-flight time reduces the radiation dose by almost a factor of three.

The trajectory is composed of the following major segments:

1) LEO to ESOI: The orbital transfer vehicle (OTV) departs on 6/24/2035 from LEO with initial mass IMLEO = 356 t, spiraling to ESOI with a fixed I_{sp} of 2200 sec. The outbound spiral segment takes 16 days and uses 120 t of propellant; after escaping from ESOI, the OTV velocity relative to Earth is 3.7 km/s.

2) Heliocentric Earth SOI to Mars SOI (MSOI): The transfer lasts 60 days and uses 41 t of propellant. The thrust direction and variable I_{sp} are maintained in the range of [2000 – 10,000] sec to minimize in-flight time.

3) Lander separation and arrival at Mars atmosphere: An arrival velocity of V_{arr} = 10 km/s is assumed with aero-braking for the Mars Lander (ML), which lands on the surface using conventional chemical propulsion. The OTV continues past Mars for rendezvous later in its orbit.

Table 10.1   General Power Plant and VASIMR® Engine System Parameters.

Reactor	exit temp T_{reac}	exit pressure P_{reac}	Pressure loss ΔP_{reac}	Mass m_{reac}
	1800 K	.4 MPa	.025 MPa	3000 kg

Generator	Enthalpy extraction ratio	Isentropic efficiency	Heat loss	Power density MW/m^3	Magnetic field B_{gen} T	Current density j_c A/m^2	Coil density ρ_c kg/m^3	Density/ Stress ρ/s_t kg/kJ
	.35	.8	.01	500	8	10^9	10^4	309
	Xe Seed Fraction	Pre-Ionizer Efficiency	Ionization Pot (eV)					
	.0001	.5	12.13					

Regenerator	Temp difference K	Heat loss	Pressure loss	Efficiency	Surface density β_{reg} kg/m^2	Heat Xfer coeff U_{reg} W/m^2/K	Density/ Stress ρ/s_t kg/kJ
	50	.01	.01	1	1	500	309

Compressor	Number of stages N_c	Isentropic efficiency $\eta_{s,c}$	Pressure loss	Radiator	Temp T_{rad} K	Pressure loss	Surface density β_{rad} kg/m^2	Emissivity ε_{rad}
	3	.85	.01		600	.01	1	.9

VASIMR®	Net power efficiency η_N	Radiator Temp T_{rad} (electronics)	Radiator Temp T_{rad} (rocket core)	Radiator Surface density β_{rad}	RF Power SS mass
	70%	300K	550K	1 kg/m^2	.4 kg/kW

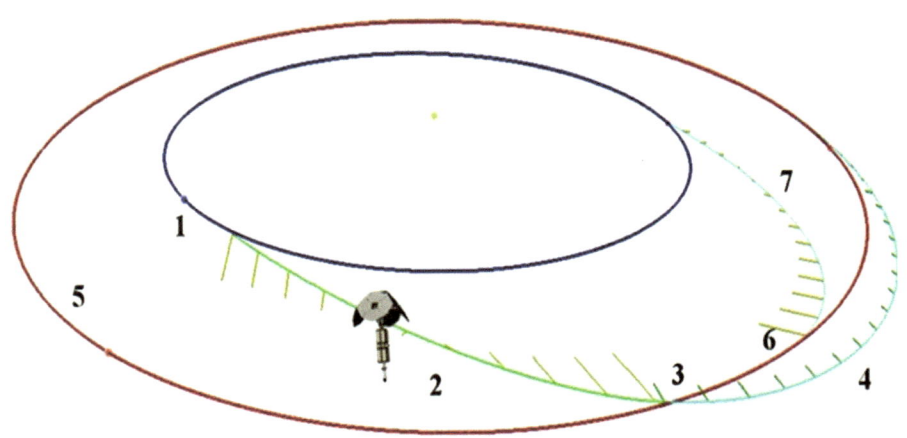

10.10. VASIMR®-NEP human mission to Mars for P = 30 MW, α = 2 kg/kW.

4) OTV rendezvous: After releasing the lander during the first Mars encounter, the OTV begins a thrust schedule that consumes 8 t of propellant over 200 days at a maximum I_{sp} = 10,000 sec, which brings the OTV within MSOI during its second encounter.

5) OTV parking: The OTV spirals down to Mars Minimum Orbit (MMO) over the course of 15 days, using 1 t of propellant at a maximum I_{sp} = 10,000 sec.

6) Mars departure: After staying on Mars for 718 days, the Crew Return Vehicle (CRV) launches from the surface and docks with the OTV. The OTV then spirals out from the one-sol orbit to MSOI in 4 days at an I_{sp} of 2200 sec, using 17 t of propellant.

7) Earth return: A heliocentric transfer from MSOI to ESOI takes 69 days and uses 27 t of propellant. The thrust direction and the variable I_{sp} is maintained in the range of [2000 – 10,000] sec to minimize the in-flight time. The OTV arrives at ESOI on 12/8/2037 with a final mass of 124 t.

Note that the nuclear VASIMR®-MHD mission described here can also be implemented with a dual I_{sp} mode, instead of the full variable. For example, using two discrete values of the specific impulse, $I_{sp,1}$ = 2200 sec and $I_{sp,2}$ = 7200 sec, increases the in-flight time for the trajectory shown in Figure 10.10 by 10 percent.

Figure 10.11 illustrates how the in-flight duration for the human mission to Mars using nuclear VASIMR®-MHD technology depends on the specific mass of the nuclear power plant and VASIMR® engines. For each value of α, there is an optimum power level that yields the shortest (out and back) in-flight transit time. The effective radiation dose, proportional to the total in-flight duration, is shown by the yellow arrow, assuming 40 g/cm² aluminum shielding for the crew while in space.

A brief study of failure scenarios for the human mission has examined unforeseen events en route, such as partial loss of reactor power and propellant loss due to leakage or plumbing failures. While these remain to be explored in greater detail, some salient features are evident. The NEP system considered here is robust in the case of failure of one, two or three of the four reactors. The partial power failure is arbitrarily assumed to occur when the vehicle has freed itself of Earth's gravity. While the failures result in an increase in the transit time, by ejecting the failed reactor(s), the mission can still be carried out at a reduced power, with the propulsion system transitioning to low I_{sp}.

A similar analysis was done in the case of the loss of propellant. For a 20 MW system operating at a specific mass of 4 kg/kW, the effect of losing up to 75 percent of the propellant remaining at the ESOI boundary was examined. In this scenario, the round-trip mission can still be completed, but at the expense of longer total flight time. The system adapts to the propellant loss by increasing the specific impulse, which consequently reduces the propellant requirement. These tradeoffs are shown in figure 10.12.

High power NEP reduces the in-flight mission time and hence the physiologically debilitating effects of prolonged interplanetary transits to Mars, including radiation exposure. Low alpha space nuclear power systems, using MHD conversion, combined with high power VASIMR® propulsion technology, offer significant advantages over the conventional NTR approach published in NASA's Mars DRA 5.0. Much work remains to be done to enable these mission capabilities, but the mass scaling and general potential of such systems are compelling, as they can lead to a significant reduction in radiation exposure to the crew, as well as inherent operational robustness in the event of unforeseen

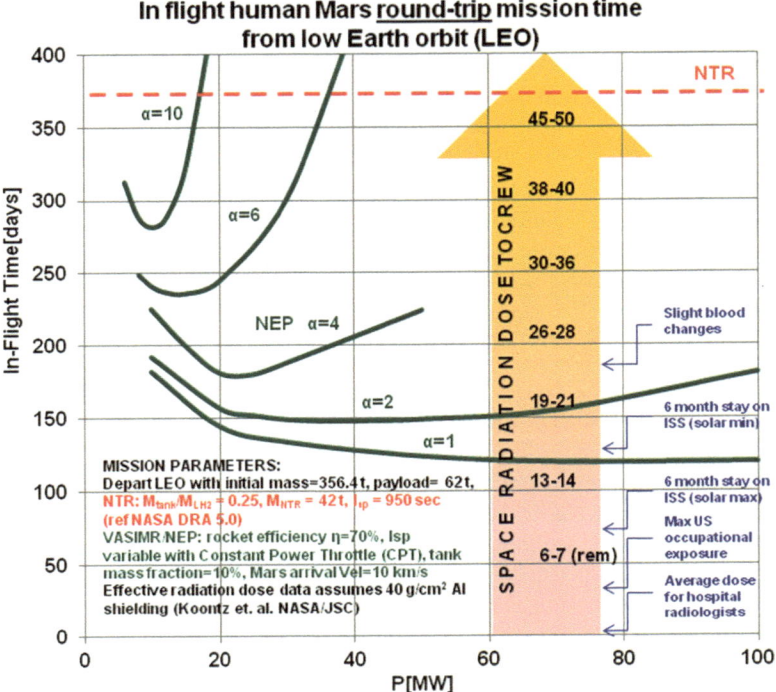

10.11. In-flight duration minima vs power for a VASIMR®-NEP human Mars mission. Increasing radiation doses are also shown for increasing interplanetary transit time. Radiation during surface time is not included.

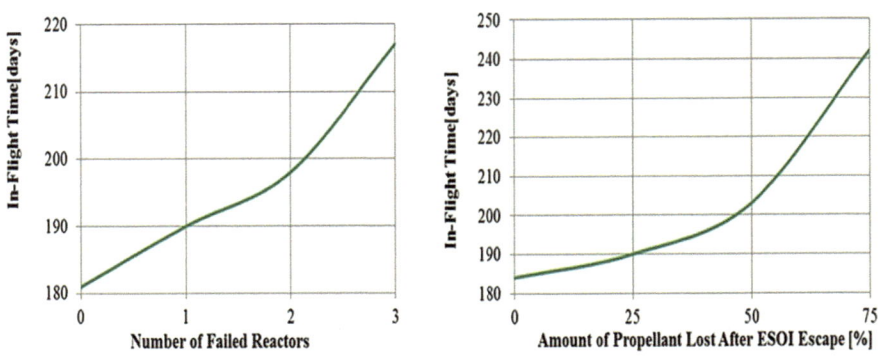

10.12. Power (left) and propellant (right) failure scenarios for a 20 MW, 4 kg/kW VASIMR®-NEP mission.

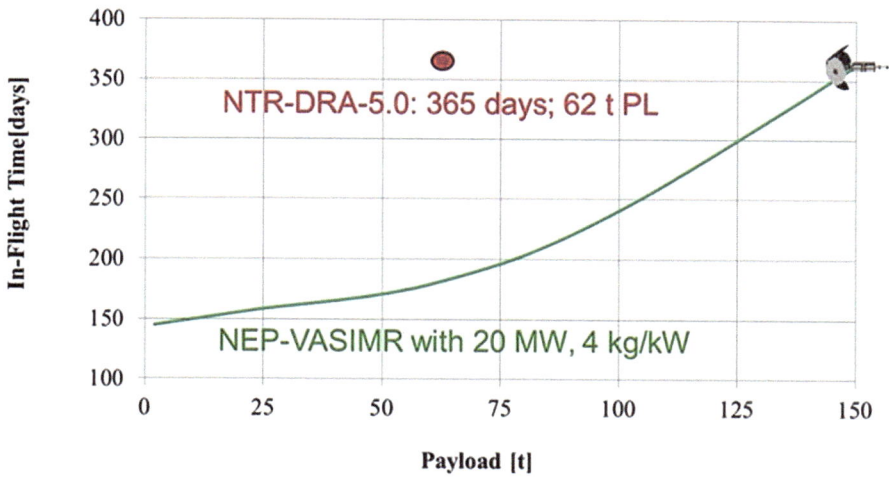

10.13. In-flight time as a function of the payload requirement for the VASIMR®-NEP human mission to Mars.

power and/or propellant system failures en route. In addition, in robotic supply missions, nuclear VASIMR®-MHD systems can deliver significantly larger payloads than their NTR or chemical counterparts.

The inherent flexibility of NEP warrants further study. For example, the in-flight time can be traded for payload and vice versa, keeping the power and propulsion system fixed, as shown in figure 10.13 for a fixed power of 20 MWe and a total specific mass 4 kg/kWe. For a long flight (as would be the case with a robotic resupply mission not involving humans), the same type of VASIMR®-NEP vehicle could be configured in a robotic cargo mode to deliver a much larger payload.

The inherent operational robustness of the NEP system is the result of the fundamental difference between NEP and the NTR or chemical options. While the latter two operate in short duration, fuel-intensive burns, the former consumes propellant more sparingly, providing thrust over virtually the entire flight.

The path to Mars and beyond, greatly facilitated by nuclear-electric power and propulsion, has a natural waypoint at the Earth's Moon. The use of the Moon as an excellent testing ground to prepare crews and systems for the journey to Mars is unquestionable. For Ad Astra, the Moon will be the perfect location to test multi-megawatt VASIMR® engines over long duration firings that will mimic a complete mission. The company plans to construct a human-tended laboratory on the surface of the Moon to carry out these tests. Multi-megawatt power can be obtained on the Moon from large solar array installations or nuclear reactors, some of which could be tested in the space environment on the Moon's surface.

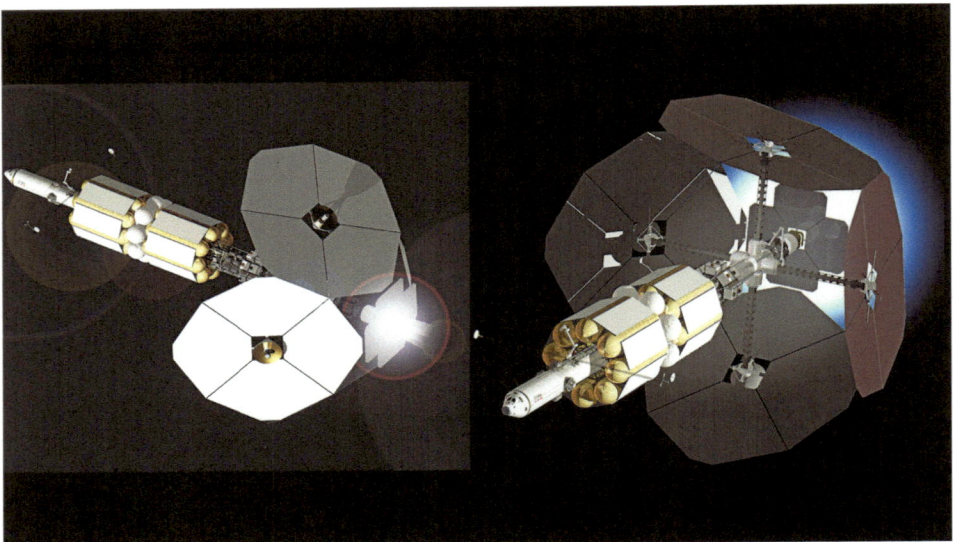

10.14. A conceptual nuclear-electric human interplanetary transport ship with VASIMR® engines as primary propulsion. The liquid hydrogen propellant is stored in a toroidal cluster of tanks to provide radiation shielding to a cylindrical crew module located on the axis. A high magnetic field superconducting radiation shield, nested inside the propellant tank cluster, provides additional shielding against galactic cosmic radiation (GCR) and secondary particles. The 4 nuclear reactors with magneto-hydrodynamic power conversion provide primary power to the engines and the rest of the ship and are located on radial booms. The reactor cores sit on top of "dish-like" high-density gamma shields and are surrounded by thermal radiating surfaces to dissipate the excess heat. High power nuclear-electric engines such as those on this ship will be fully tested at Ad Astra's future lunar test facility on the surface of the Moon.

An economically sustainable human exploration of Mars and points beyond in the solar system must not be approached single handedly by one nation. It can only be achieved with a concerted and well-balanced international effort that fosters interdependence, but also healthy competition. A robust human exploration of our solar system will not be possible without the development of advanced nuclear-electric power and propulsion. Global commerce, business and entrepreneurial activities must extend into space to establish a robust space-faring economy for the planet and are essential to assure the permanence and resilience of human activities outside of the Earth and, indeed, to ensure our survival as a species. These activities must not be delayed or hindered by unreasonable restrictions in technology transfer. As has occurred with the International Space Station, humans of all nations are able to work together in space as citizens of a planet. More and more citizens of all nations need to have the opportunity to fly in space and the more technologically advanced countries have the duty to help this process of space democratization. Perhaps the elusive path to world peace actually lies in the heavens.

REFERENCES

1. *Rapid Mars Transits with Exhaust Modulated Plasma Propulsion;* NASA Technical Paper 3539, May 1995.
2. Stuhlinger, E., *Ion Propulsion for Space Flight,* McGraw-Hill, 1964.
3. Fast and Robust Human Missions to Mars with Advanced Nuclear Electric Power and VASIMR® Propulsion. Franklin R. Chang Díaz[1], Mark D. Carter[1], Timothy W. Glover[1], Andrew V. Ilin[1], Christopher S. Olsen[1], Jared P. Squire[1], Ron J. Litchford[2], Nobuhiro Harada[3], Steven L. Koontz[4],

   1: Ad Astra Rocket Company, 141 W. Bay Area Blvd., Webster, TX 77598. 2: NASA Marshall Space Flight Center, Huntsville, AL 35812. 3: Nagaoka University of Technology, Nagaoka 940-2188, Japan. 4: NASA Johnson Space Center, Houston, TX 77058. Proceedings of Nuclear and Emerging Technologies for Space 2013. Albuquerque, NM, February 25-28, 2013. Paper 6777

Index

© Springer International Publishing Switzerland 2017
F. Chang Díaz, E. Seedhouse, *To Mars and Beyond, Fast!*, Springer Praxis Books,
DOI 10.1007/978-3-319-22918-8